T0201078

NANOCHROMATOGRAPHY AND NANOCAPILLARY ELECTROPHORESIS

NANOCHROMATOGRAPHY AND NANOCAPILLARY ELECTROPHORESIS
Pharmaceutical and Environmental Analyses

IMRAN ALI

HASSAN Y. ABOUL-ENEIN

VINOD K. GUPTA

A JOHN WILEY & SONS, INC., PUBLICATION

Copyright © 2009 by John Wiley & Sons, Inc. All rights reserved

Published by John Wiley & Sons, Inc., Hoboken, New Jersey

Published simultaneously in Canada

No part of this publication may be reproduced, stored in a retrieval system, or transmitted in any form
or by any means, electronic, mechanical, photocopying, recording, scanning, or otherwise, except as
permitted under Section 107 or 108 of the 1976 United States Copyright Act, without either the prior written
permission of the Publisher, or authorization through payment of the appropriate per copy fee to the
Copyright Clearance Center, Inc., 222 Rosewood Drive, Danvers, MA 01923, (978) 750-8400, fax (978)
750-4470, or on the web at www.copyright.com. Requests to the Publisher for permission should be
addressed to the Permissions Department, John Wiley & Sons, Inc., 111 River Street, Hoboken, NJ
07030, (201) 748-6011, fax (201) 748-6008, or online at http://www.wiley.com/go/permission.

Limit of Liability/Disclaimer of Warranty: While the publisher and author have used their best efforts
in preparing this book, they make no representations or warranties with respect to the accuracy or
completeness of the contents of this book and specifically disclaim any implied warranties of
merchantability or fitness for a particular purpose. No warranty may be created or extended by sales
representatives or written sales materials. The advice and strategies contained herein may not be suitable
for your situation. You should consult with a professional where appropriate. Neither the publisher nor
author shall be liable for any loss of profit or any other commercial damages, including but not limited
to special, incidental, consequential, or other damages.

For general information on our other products and services or for technical support, please contact our
Customer Care Department within the United States at (800) 762-2974, outside the United States at
(317) 572-3993 or fax (317) 572-4002.

Wiley also publishes its books in a variety of electronic formats. Some content that appears in print may not
be available in electronic formats. For more information about Wiley products, visit our web site at www.
wiley.com.

Library of Congress Cataloging-in-Publication Data:

Ali, Imran.
    Nano chromatography and nano capillary electrophoresis: pharmaceutical and environmental
analyses / Imran Ali, Hassan Y. Aboul-Enein, Vinod K. Gupta.
        p.; cm.
    Includes bibliographical references and index.
    ISBN 978-0-470-17851-5 (cloth)
1. Chromatographic analysis.   2.   Capillary electrophoresis.   3.   Drugs—Analysis.
4.   Pollutants—Analysis.   5.   Environmental pollutants—Analysis.   6.   Nanoparticles—Analysis.
I. Aboul-Enein, Hassan Y. II. Gupta, Vinod Kumar, 1953– III. Title.
    [DNLM: 1.   Nanotechnology—methods.   2.   Electrophoresis, Capillary—methods.
3.   Environmental Pollutants—analysis.   4.   Pharmaceutical Preparations—analysis.
QT 36.5 A398n 2009]

QP519.9.C47A45 2009
543′.8—dc22

                                                                                2008033275

Printed in the United States of America

10 9 8 7 6 5 4 3 2 1

To the memories of my late parents:
Basheer Ahmed and Mehmudan Begum.
I. A.

To my wife, Nagla, whose love and devotion
have been an inspiration to me.
H. Y. A.-E.

To the memory of my late father Sri Jeewan Lal Gupta
who has been my mentor all through.
V. K. G.

CONTENTS

6 Nano-High Performance Liquid Chromatography   145

7 Nanocapillary Electrochromatography and Nanomicellar Electrokinetic Chromatography   167

PREFACE

With the advancement of science and technology, nanoanalysis is becoming more important because scientists, academicians, and regulatory authorities are asking for data detection at the nanogram level. This need is especially pressing in genomics, proteomics, and drug designing and development programs. Our search of the literature and experience dictates that chip-based analytical techniques viz. nanoliquid chromatography and nanocapillary electrophoresis are the best choices for such types of applications. Therefore, the separation and identification of many compounds in biological and environmental matrices at nanolevels are gaining importance day by day. In view of these facts, this book deals with the fabrication of microfluidic devices, instrumentation, detection, sample preparation, and applications of nanoliquid chromatography and nanocapillary electrophoresis techniques for analyses at the nanogram level. This book describes analyses at the nanogram-per-level in detail with main emphasis on experimental methodologies. Moreover, we explain optimization strategies, helpful to design future experiments in this area. This book is important and unique because it is one of the first texts to be published that addresses chip-based nanoanalyses. This is a useful reference for scientists, researchers, academicians, and graduate students working in the field of nanoanalyses. Uniquely, this book also fulfills the requirements of regulatory authorities for formulating regulations and legislations to control the dosage of drug and contaminant exposure to the environment.

IMRAN ALI
HASSAN Y. ABOUL-ENEIN
VINOD K. GUPTA

New Delhi, India
Cairo, Egypt
Roorkee, India
January 2009

ACKNOWLEDGMENTS

It was indeed a difficult task for me to complete this book but the extreme help and cooperation of my wife, Seema Imran, made it reality. Thanks also to my dear son, Al-Arsh Basheer Baichain, who has given me freshness and fragrance continuously during the completion of this difficult job. I acknowledge my other family members, relatives, and laboratory staff who have helped me directly and indirectly during this period.

My sincere thanks to Professor Kishwar Saleem and Professor Tabrez A. Khan, Department of Chemistry, JMI, New Delhi, who helped me to complete this book. Moreover, their constant and continuous moral support was the biggest help and a memorable event in my life. Finally, John Wiley & Sons is also acknowledged for providing financial assistance to complete this work. Reader comments are welcome at my email address: drimran_ali@yahoo.com.

I. A.

I am grateful to my wife, Nagla El-Mojadaddy, for her forbearance and support throughout the preparation of this book and it is to her that I extend my deepest gratitude. I acknowledge her patience and tolerance during the preparation of this book.

Thanks are extended to the editorial staff of John Wiley & Sons for their assistance in publishing this book.

H. Y. A.-E.

I acknowledge the love of my mother, Smt. Kiran Devi, my wife, Prerna Gupta, my son, Rajat Shikhar Gupta, my daughter, Vartika Gupta, and granddaughter, Shiriya Gupta, who have been the sources of inspiration and given me freshness and fragrance continuously during the completion of this job. Acknowledgments are also due to my other family members, relatives, and laboratory staff who have helped me directly and indirectly during this period.

V. K. G.

CHAPTER 1

INTRODUCTION

1.1. NANOANALYSES

Separation science is the backbone of biological, biotechnological, chemical, pharmaceutical, environmental, geochemical, and agricultural disciplines and, hence, analysis is an integral part of these fields [1,2]. It is a well-known fact that almost all drugs work at low concentrations in biological systems and to carry out pharmacokinetic and pharmacodynamic studies analytical techniques should be capable of detecting drugs and pharmaceuticals at nano or lower detection limits. In spite of the curative properties of drugs, some drugs also have side effects even at low concentrations, that is, ranging from nanogram to femtogram levels [3,4]. Therefore, in the absence of techniques capable of detecting at the nanolevel we assume the absence of drug residues in the body, while these can have some bioreactions and side effects. Besides, the concentrations of some species, such as hormones, RNA, DNA, antibodies, and other proteins are very low, and require analytical techniques with low detection limits. In addition, detection of drugs at a lower concentration is required in the plasma of infants because of the availability of only limited amounts of blood samples. In addition, for some biological fluids, such as cerebrospinal fluids, only small volumes are available for sampling. Besides, high throughput screening (HTS) and drug discovery (combinatorial

Nanochromatography and Nanocapillary Electrophoresis. By Ali, Aboul-Enein, and Gupta
Copyright © 2009 John Wiley & Sons, Inc.

1

chemistry) need low level analyses. Moreover, recent advancements in proteomics and genomics compel scientists to develop nanoanalytical techniques.

Similarly, many xenobiotics, such as pesticides, polynuclear aromatic hydrocarbons (PAHs), polychlorinated biphenyls (PCBs), polybrominated diphenyl ethers (PBDEs), plasticizers, phenols, and some other drug residues, are also toxic even at trace levels present in the earth's ecosystem [5–7]. Without analytical techniques capable of detecting them at nanolevels, we assume the absence of these pollutants in the environment, while these notorious pollutants accumulate in our body tissues resulting in various diseases and side effects such as carcinogenesis and failure of many vital body organs including the kidney, liver, and heart [8–11]. Under such situations, it is essential to have analytical techniques that can detect drugs, pharmaceuticals, and xenobiotics in biological and environmental samples at very low concentrations.

Apart from the above requirements of nanoanalyses, the need of detection at the nanolevel is also increasing continuously in dosage formulations, food products, and other chemical and biotechnology industries. Briefly, nowadays, nanoanalysis is becoming more important and scientists and regulatory authorities are asking for data on detection at the nanogram level. Among various analytical techniques, chromatography and capillary electrophoresis are the choice of scientists, academicians, and clinicians as these techniques can analyze samples of low volume or having poor ingredient concentrations. Some attempts have been made to develop nanochromatography and nanocapillary electrophoresis, which some workers call the micrototal-analysis system (μ-TAS). In this book we have replaced μ-TAS with the phrase nanoanalysis as the techniques are capable of dealing with nano amounts of samples, with detection at the nanogram level.

1.2. DEFINITION OF NANOCHROMATOGRAPHY AND NANOCAPILLARY ELECTROPHORESIS

Karlsson and Novotny [12] introduced the concept of nanoliquid chromatography in 1988. The authors reported that the separation efficiency of slurry packed liquid chromatography microcolumns (44 μm, id) was very high. Since then, many advance have been reported in this modality of chromatography and it has been used as a complementary and/or competitive separation method to conventional chromatography. Unfortunately, to date no correct and specific definition of this technique has been proposed, probably due to the use of varied column sizes (10 to 140 μm). Some definitions of nanoliquid chromatography are found in the literature based on column diameter and mobile phase flow rates [13–15]. It has been reported that when the chromatographic separation is carried out in capillary columns of 10 to 100 μm internal diameter, the modality is called nanoliquid chromatography, whereas when capillaries of 100 to 500 μm internal diameter are used the technique is called capillary liquid chromatography [16]. On the other hand, some workers define nanoliquid chromatography as the chromatographic modality having a mobile flow rate of nanomilliliters per minute. However, no one has considered the detection aspect of this type of chromatography, which is very important in analytical science.

Our intention is to achieve nanolevel detection irrespective of the experimental condition as detection is the final aim in separation science. Moreover, nanolevel detection is required for low amounts of sample or samples having poor ingredients. Therefore, if the technique is capable of detecting at low or nanolevels one can analyze low volume samples or samples having poor concentrations. Therefore, detection is the most important issue in nanochromatography. Based on these requirements and logics we have defined nanoliquid chromatography more accurately and scientifically. Keeping all these facts in mind, nanochromatography may be defined as "a modality of chromatography involving samples in nanoliters, mobile phase flow in nanomilliliters per minute, with detection at the nanogram per milliliter level." This definition is a complete one and all the requirements can be fulfilled on chip-based chromatography. Therefore, a true and complete nanochromatography is only possible on a chip, which is called lab-on-chip chromatography but we simply named it nanochromatograpy (NC) broadly. In the case of liquid chromatography it may be called nanoliquid chromatography and abbreviated as NLC. The same definition is also true and complete for capillary electrophoresis, which is also possible on chips, and we have termed it nanocapillary electrophoresis (NCE).

1.3. NANOCHROMATOGRAPHY AND NANOCAPILLARY ELECTROPHORESIS

Recently, the word *nano* has become a trend in science and technology and some of us think that it is the new generation but, as mentioned above, its root is about 22 years old. NLC and NCE are gaining importance day by day. They are very useful and effective tools for samples of low quantities or having low concentrations of the analytes. Columns of low internal diameter are ideal for use in NLC and NCE, especially with detectors requiring very low flow rates, such as electrospray liquid chromatography/mass spectroscopy (LC/MS). Besides, these columns offer high sensitivity due to their low dispersion characteristics. The microfluidic systems are more or less capable of replacing conventional "macro" systems for many applications in the life sciences. The working principle of NLC and NCE are the same as in conventional LC and CE. However, miniaturization offers many advantages over the conventional methods, including:

- Usefulness and effectiveness for samples of low volume or having extremely low concentrations of ingredients.
- Significantly reducing solvent consumption and subsequent waste production.
- Inexpensive due to low consumption of solvent, electricity, and operational time.
- Good potential portability due to a system size reduction.
- High sensitivity, speed, and reproducibilities.
- Narrowing the peak width of chromatogram/electropherogram due to better separation efficiency.
- Low mobile phase pressure in NLC.
- Simultaneous mass separation on chips.
- Good hyphenation of detectors requiring mobile phase flow.

NCE is a relatively new development in separation science, especially in proteomics and genomics. In the last two decades NCE has gained increasing importance, as can be seen from a good number of publications [17–20]. In addition to the above advantages, NCE is a suitable technique for samples that may be difficult to separate by NLC as the principles of separation are entirely different. Lower detection limits of NCE lead to the possibility of separating and characterizing small quantities of materials. Moreover, the enzymatic reactions for analytical purposes can be conducted within the capillary.

1.4. FABRICATION OF MICRODEVICES

The fabrication of microdevices in NLC and NCE is controlled by highly developed batch-processing techniques of integrated circuits. The microelectronic technology can be exploited for microflow systems with functions different from those of integrated circuits. Special processes and materials are needed, called micromachining techniques [21]. Following developments in miniaturization and integration of electronic devices, the potential of a similar revolution also emerged for mechanical and later on fluidic devices, which led to MEMS (microelectromechanical systems) and μ-TAS (micrototal analysis systems) [22]. μ-TAS is one of the fastest growing areas covering microfluidics, material science, analytical chemistry, and biotechnology. Microfluidics refers to the science and technology dealing with minute amounts of fluids (micro-, nano-, and picoliters). Lab-on-a-chip is a miniaturization and integration of complete functionality of a chemistry or biology lab, for example, preparation, reactions, separation, and detection, onto a single chip. μ-TAS is a development involving reduced size, low power, sample, reagent and manufacture requirements and operating costs. Besides, μ-TAS can perform better services in terms of speed, throughput, mass sensitivity, and automation.

Microfluidics is the key to NLC and NCE, the miniaturized microfluidic system that can automatically carry out all the necessary functions to transform chemical information into electronic information. The first μ-TAS device was developed by Terry et al. [23] for gas chromatography, which did not gain popularity at that time, probably due to poorly developed microfluidic devices. In 1990, Manz et al. [24] introduced the concept of μ-TAS. Nowadays, μ-TAS is a popular development in various disciplines and has been reviewed by Manz and coworkers [25–28].

The development era of microdevices was from 1970 to 1980 in the silicon microprocessors industry. Interested readers can consult textbooks on this subject [29–31]. Kim et al. [32] presented a molding protocol for patterning network channels by contacting a substrate and a patterned elastomeric master. Later on Mrksich and Whitesides [33] used microcontact printing to pattern structures of self-assembled monolayers (SAMs) on the submicrometer scale. In 1996, Zhao et al. [34] used microtransfer molding for fast fabrication of organic polymers and ceramics in three dimensions using a layer-by-layer structuring. Park and Madou [35] designed a three-dimensional electrode useful for a high throughput dielectrophoretic separation/concentration/filtration system. Clicq et al. [36] used reversed phase chromatography on a rectangular glass chip coated with C_8 silica gel. Many other attempts have been made by various workers toward fabrication of microdevices [37–42]. The design of microdevices is a very important issue, especially in NLC and NCE. Bousse et al. [43] reported a microfabricated electrokinetic device having loading and separation channels. Manz and Becker [44] developed a design in NCE useful for effective working of the system. The introduction of stationary phases in NLC is quite difficult; however, some workers have attempted this [45–48]. The stationary phases were introduced into a microchannel by coating of the inner surface of the capillary or by packing the channels or in situ polymerization of continuous beds. Some microfluidic-NLC chips have been commercialized by a few manufacturers.

1.5. DEVELOPMENTS IN NANOANALYSES

Karlsson and Novotny [12] introduced the nanoliquid chromatography (nano-LC) concept and since then it has been proposed as a complementary and/or competitive separation method to conventional liquid chromatography. Nano detection is being achieved by modified LC systems but it is still in its development stage [49]. Generally, column internal diameters in the range of 25 to 100 μm are considered under nanochromatographic systems. But the miniaturization and integration of electronic devices led to the development of microelectromechanical systems (MEMS) and micrototal analysis systems (μ-TAS), which are components of chip-based analytical machines as MEMS devices are faster, selective, sensitive, and economic in nature. μ-TAS is a fast-growing area in separation science. Of course, lab-on-chip systems are not yet fully developed but demand for them is increasing due to their economic, fast, and portable capabilities. A complete analysis can be carried out in seconds by using 1 to 5 mL of solvent. Moreover, the machines are very compact and can be carried to sites for environmental analyses.

Some important developments and advances in NLC and NCE are described in papers by Manz and coworkers [25–28]. These authors [50] presented a miniaturized open-tubular liquid chromatograph on a silicon wafer with a 5 × 5 mm silicon chip containing a conductometric detector, connected to an off-chip conventional LC pump and valves to perform high-pressure liquid chromatography. Manz et al. [24] proposed a μ-TAS concept in which silicon chip analyzers having sample pretreatment, separation, and detection played a fundamental role. In the early stage the intention of miniaturization was to increase analytical performance of the device rather than reduction in size. It was also realized that miniaturization provided the advantages of a smaller consumption of reagent and time. Moreover, μ-TAS provided an integration of separation techniques, as it is capable of performing sample handling, analysis, and detection on a single chip. Later, Jacobson et al. [51] used a microchip of fused quartz to separate complexed metal ions in polyacrylamide-modified channels. In the same year Ocvirk et al. [45] developed NLC with a split injector, a packed small-bore column, a frit, and an optical detector cell onto a silicon chip. Moore et al. [52] reported chip-based micellar electrokinetic capillary chromatography (MECC) of neutral dyes. Similarly, von Heeren et al. [53] analyzed biological samples on MECC fabricated onto a chip. Penrose et al. [54] reported a centrifugal chromatograph for reversed-phase separations. Recently, Zeng et al. [55] reported the chiral separation of amino acids on polydimethylsiloxane (PDMS) chips.

Chip-based technology is successful in NCE over NLC due to difficulties of integrating on chip pumps, injectors, mechanical valves, and the lack of easy flow control [56]. Besides, it is also difficult to install a stationary phase in NLC and sealing of the microchannels, which should be perfect for the mobile phase to flow appropriately through the stationary phase. In 1992, Manz et al. [57,58] introduced the first NCE integrated silicon and glass chips. The authors described the concept of μ-TAS in NCE by integrating injection, separation, and detection on the chip. Furthermore, some advancement has been reported in this direction by the same workers [59–62]. Woolley et al. [63] developed capillary array electrophoresis (CAE) for the analysis of different DNA samples. Since then much work has been done in this area and some quite good papers are available, which have been considered for preparing this book. Other developments in NCE have been reported from time to time and can be used for the analyses of different compounds [64–83]. Another development in NCE was the enantiomeric resolution of amines on chip-based NCE by Cong and Hauser [84]. Reviews have been published describing various aspects of microdevices [25–28,85–91].

1.6. DATA INTEGRATION

As mentioned above, the basic principle of NLC is the same as for conventional techniques. The separation is identified and characterized by measuring retention times, capacity, separation, and resolution factors. Therefore, it is necessary to explain the chromatographic terms and symbols by which the chromatographic speciation can be understood and explained. Some of the important terms and equations of the chromatographic separations are discussed below. The chromatographic separations are characterized by retention (k), separation (α), and resolution factors (Rs). The values of these parameters can be calculated by the following standard equations [92].

$$k = (t_r - t_0)/t_0 \qquad (1.1)$$

$$\alpha = k_1/k_2 \qquad (1.2)$$

$$Rs = 2\,\Delta t_r/(w_1 + w_2) \qquad (1.3)$$

where t_r and t_0 are the retention time of the separated species and dead time (solvent front) of the column in minutes/seconds, respectively. Δt, w_1, and w_2 are the difference in the retention times of two peaks of the separated species, the base width of peak 1 and peak 2, respectively. If individual values of α and Rs are 1 or greater the separation is supposed to be complete. If the individual values of these parameters are lower than 1 the separation is understood as partial or incomplete.

The number of a theoretical plate (N) characterizes the quality of a column/chip. The larger N is, the more complicated the sample mixture that can be separated with the column. The value of N can be calculated from the following equations:

$$N = 16(tr/w)^2 \qquad (1.4)$$

or

$$N = 5.54[(tr)/w_{1/2}]^2 \qquad (1.5)$$

where t_r, w, and $w_{1/2}$ are the retention times in minutes or seconds of the peak, peak width at the base, and at half the height of the peak, respectively. The height equivalent to a theoretical plate [HETP (h)] is a section of a column or chip in which the mobile and the stationary phases are in equilibrium. Since a large number of theoretical plates are desired, h should be as small as possible. Naturally, there are no real plates in a column or chip. The concept of a theoretical plate is a variable and its value depends on particle size, flow velocity, mobile phase (viscosity), and especially on the quality of the packing h can be calculated from the following equation:

$$h = L/N \qquad (1.6)$$

where L is the length of the column used.

The mechanism of separation in NCE is based on the difference in the electrophoretic mobility of the separated species. Under NCE conditions, the migration of the separated species is controlled by the sum of the intrinsic electrophoretic mobility (μ_{ep}) and the electroosmotic mobility (μ_{eo}), due to the action of electroosmotic flow (EOF). The observed mobility (μ_{obs}) of the species is related to μ_{eo} and μ_{ep} by the following equation:

$$\mu_{obs} = E(\mu_{eo} + \mu_{ep}) \qquad (1.7)$$

where E is the applied voltage (kV).

The simplest way to characterize the separation of two components, the resolution factor (Rs), is to divide the difference in the migration times by the average peak width as follows:

$$Rs = 2(t_2 - t_1)/(w_1 + w_2) \qquad (1.8)$$

where t_1, t_2, w_1, and w_2 are migration times of peak 1 and peak 2, and the widths of peak 1 and peak 2, respectively.

The value of the separation factor may be correlated with μ_{app} and μ_{ave} by the following equation:

$$Rs = (1/4)(\Delta\mu_{app}/\mu_{ave})N^{1/2} \tag{1.9}$$

where μ_{app} is the apparent mobility of two separated species and μ_{ave} is the average mobility of the separated moieties. Using Equation 1.9 permits independent assessment of two factors that affect separation, selectivity and efficiency. The selectivity is reflected in mobility of the analytes while the efficiency of the separation process is indicated by N. Another expression for N is derived from the following equation:

$$N = 5.54(L/w_{1/2})^2 \tag{1.10}$$

where L and $w_{1/2}$ are the capillary length and peak width at half height, respectively. Here it is important to point out that it is misleading to discuss theoretical plates in NCE and it is simply a carryover from chromatographic theory. The theoretical plate in NCE is merely a convenient concept to describe analyte peak shape and to assess the factors that affect separation.

HETP may be considered as the function of the capillary occupied by the analyte and more practical to measure separation efficiency in comparison to N. σ^2_{tot} is affected not only by diffusion but also by differences in the mobilities, heating of the capillary in joules, and interaction of the analytes with the capillary wall, and hence σ^2_{tot} can be represented as shown in Equation 1.11.

$$\sigma^2_{tot} = \sigma^2_{diff} + \sigma^2_{T} + \sigma^2_{int} + \sigma^2_{wall} + \sigma^2_{Electos} + \sigma^2_{Electmig} + \sigma^2_{Sorp} + \sigma^2_{Oth} \tag{1.11}$$

where the values represent the square roots of the standard deviations of total, diffusion, heat, injection, wall, electroosmosis, electromigration, sorption, and other phenomena, respectively.

Practically, velocity (v) and observed mobility (μ_{obs}) of the analyte, electrophoretic mobility (μ_{ep}), and electroosmotic mobility (μ_{eo}) can be calculated by the following equations.

$$v = \mu_{obs}E \tag{1.12}$$

where E is the strength of the electric field in V/cm.

E is described by the following equation:

$$E = V/Lt \tag{1.13}$$

where Lt is the total length of the capillary or channel.

By putting the value of Equation 1.13 into Equation 1.12

$$v = (\mu_{obs}V)/Lt$$

or

$$\mu_{obs} = (vLt)/V \tag{1.14}$$

We know that the velocity (v) of the analyte equals distance divided by time and, hence, velocity in NCE may be written by the following equation:

$$v = Ld/td \quad (\text{cm sec}^{-1}) \tag{1.15}$$

where Ld and td are the distance of the capillary or channel up to the detector point and the time of migration of the analyte up to the detector, respectively.

By putting the value of v from Equation 1.15 into Equation 1.14

$$\mu_{obs} = (Ld/td) \times (Lt/V) \tag{1.16}$$

or

$$\mu_{ep} + \mu_{eo} = (Ld/td) \times (Lt/V) \tag{1.17}$$

For benzene the value of μ_{ep} is zero and hence:

$$\mu_{eo} = Ld/td \times Lt/V \quad (\text{cm}^2\text{V}^{-1}\text{sec}^{-1}) \tag{1.18}$$

Equation 1.17 may be written in the following form:

$$\mu_{ep} = Ld/td \times Lt/V - \mu_{eo} \quad (\text{cm}^2\text{V}^{-1}\text{sec}^{-1}) \tag{1.19}$$

By using all these equations, we can calculate all parameters of NCE required in any analysis.

1.7. PROTOCOL OF NANOANALYSES

Of course, conventional chromatographic and capillary electrophoretic methods are popular for the analyses of drugs, pharmaceuticals, and pollutants in biological and environmental matrices. In the last few decades, new developments in nanoanalyses have emerged in the literature due to their high separation efficacy, speed, inexpensive running cost, and small volume requirement, which are major

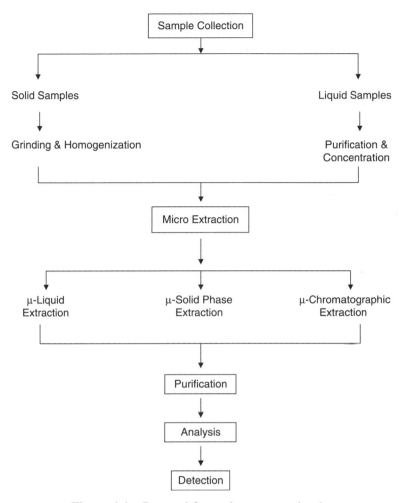

Figure 1.1   Protocol for analyses at nanolevels.

trends in analytical science. The number of publications in separation science using this development have increased. Use of these techniques is difficult because it involves micro or nano amounts of samples. A well-trained operator is required for handling NLC and NCE. A good protocol is always useful to work with these modalities of chromatography and electrophoresis. The working protocol for NCL and NCE is shown in Fig. 1.1 and a careful perusal of this figure indicates that sample preparation is a very important issue in these modalities, especially in unknown matrices. Our experience dictates that online sample preparation devices are required to avoid any error of methods due to small sample quantities. Optimization and other issues related to nanoanalyses are discussed in later chapters of the book.

1.8. SCOPE OF THE BOOK

This book explains the importance and applications of nanoanalytical techniques with special emphasis on separation of drugs, pharmaceuticals, and xenobiotics in biological and environmental matrices. A thorough search of the literature was carried out and many papers are available on nanoanalyses. As per the definition of NLC and NCE, microchip-based techniques are true nanoanalytical methods but only a few papers are available using these modalities, which have been included in this book. Some publications are also available on chromatography and capillary electrophoresis with columns with internal diameters ranging from 25 to 100 μm and flow rates of 25 to 4,000 nL/minute. In addition to this, a few papers describe detection at the nanogram level by using conventional LC and CE methods. These papers describing separations on microbore columns and normal columns with detection at the nanogram per milliliter level have been included in this book to make it more useful for the reader. The intention behind this is to cover all the analyses dealing with nanolevel detection, which is the last destination of nano world separation science. Arbitrarily, publications having 0.5 ng/mL (500 ng/L) or lower as the detection limits for the analyte(s) were considered as nanoanalyses and included in this book. However, the main emphasis has been given to the instrumentation and analyses on chip-based NCE and NLC.

1.9. CONCLUSION

The development of microdevices led to a new area in separation science called μ-TAS or lab-on-chip technology. But we have given a new term, nanoanalysis, to this type of analysis. The chip-based microfluidic devices are well recognized and used in many areas of biological sciences [93]. This book discusses fabrication, instrumentation, detection, nanoliquid-chromatography (NLC), and nanocapillary electrophoresis (NCE) since these techniques deal with nano amounts of sample, flow rate, and detection as well. We hope that this text will be useful for graduate students, researchers (biological and environmental sciences), and professionals of pharmaceutical, agrochemical, and other chemical industries.

REFERENCES

1. A. Connors, *A textbook of pharmaceutical analysis*, New York: Wiley & Sons (1982).
2. D.A. Skoog, *Fundamentals of analytical chemistry*, 8th ed., Belmont, CA: Thomson-Brooks/Cole (2004).

3. D.M. Bare, *Med. Lab. Obs.*, **36**, 10 (2004).
4. W. Tse, *J. Am. Med. Dir. Assoc.*, **7**, 12 (2006).
5. V. Gaston, *International regulatory aspects for chemicals*, Vol. I, New York: CRC Press (1979).
6. D.Z. John, *Handbook of drinking water quality: Standards and controls*, New York: Van Nostrand Reinhold (1990).
7. *Toxic Substance Control Act*, US EPA, III, 344 (1984).
8. E.J. Ariens, *Trends. Pharmacol. Sci.*, **14**, 68 (1993).
9. J.C. Leffingwell, *Chirality and bioactivity I: Pharmacology, Leffingwell, Reports*, **3**, (2003) 1, www.leffingwell.com.
10. I. Ali, H.Y. Aboul-Enein, *Instrumental methods in metal ions speciation: Chromatography, capillary electrophoresis and electrochemistry*, New York: Taylor & Francis (2006).
11. I. Ali, H.Y. Aboul-Enein, *Chiral pollutants: Distribution, toxicity and analysis by chromatography and capillary electrophoresis*, Chichester, UK: John Wiley & Sons (2004).
12. K.E. Karlsson, M. Novotny, *Anal Chem.*, **60**, 1662 (1988).
13. J.P.C. Vissers, H.A. Claessens, C.A. Cramers, *J. Chromatogr. A*, **779**, 1 (1997).
14. J.P. Chervil, M. Ursem, J.P. Salzmann, *Anal. Chem.*, **68**, 1507 (1996).
15. Y. Saito, K. Jinno, T. Greibrokk, *J. Sep. Sci.*, **27**, 1379 (2004).
16. J.H. Hernandez-Borges, Z. Aturki, A. Rocco, S. Fanali, *J. Sep. Sci.*, **30**, 1589 (2007).
17. J. Khandurina, A. Guttman, *Curr. Opin. Chem. Biol.*, **7**, 595 (2003).
18. G. Rozing, *LC GC Europe*, **16**, 14 (2003).
19. C. Legido-Quigley, N.D. Marlin, V. Melin, A. Manz, N.W. Smith, *Electrophoresis*, **24**, 917 (2003).
20. Z. Deyl, F. Svec (Eds.), *Capillary electrochromatography*, Amsterdam: Elsevier (2001).
21. S. Büttgenbach, In *Non-linear electromagnetic systems* (Eds: V. Kose, J. Sievert), Amsterdam: IOS Press (1998).
22. Q. He, *Integrated nano liquid chromatography system on a chip*, Ph.D. thesis, California Institute of Technology, Pasadena, CA, June 21 (2005).
23. S.C. Terry, J.H. Jerman, J.B. Angell, *IEEE Trans. Electron. Devices*, **26**, 1880 (1979).
24. A. Manz, N. Graber, H.M. Widmer, *Sensors and Actuators*, **B1**, 244 (1990).
25. D.R. Reyes, D. Iossifidis, P.A. Auroux, A. Manz, *Anal. Chem.*, **74**, 2623 (2002).
26. P.A. Auroux, D. Iossifidis, D.R. Reyes, A. Manz, *Anal. Chem.*, **74**, 637 (2002).
27. T. Vilkner, D. Janasek, A. Manz, *Anal. Chem.*, **76**, 3373 (2004).
28. P.S. Dittrich, K. Tachikawa, A. Manz, *Anal. Chem.*, **78**, 3887 (2006).

29. P. Rai-Choudhury, *Handbook of microlithography, micromachining, and microfabrication*, 1st ed., Vol. 2, Bellingham: SPIE Press (1997).

30. P. Rai-Choudhury, *Handbook of microlithography, micromachining, and microfabrication*, 1st ed., Vol. 1, Bellingham: SPIE Press (1997).

31. M.J. Madou, *Fundamentals of microfabrication: The science of miniaturization*, 2nd ed., Boca Raton: CRC Press (2002).

32. E. Kim, Y.N. Xia, G.M. Whitesides, *Nature*, **376**, 581 (1995).

33. M. Mrksich, G.M. Whitesides, *Trends Biotechnol.*, **13**, 228 (1995).

34. X.M Zhao, Y.N. Xia, G.M. Whitesides, *Adv. Mater.*, **8**, 837 (1996).

35. B.Y. Park, M.J. Madou, *Electrophoresis*, **26**, 3745 (2006).

36. D. Clicq, N. Vervoort, R. Vounckx, H. Ottevaere, J. Buijs, C. Gooijer, F. Ariese, G.V. Baron, G. Desmet, *J. Chromatogr. A*, **979**, 33 (2002).

37. K.S. Johnson, K.K. Berggren, A. Black, C.T. Black, A.P. Chu, N.H. Dekker, D.C. Ralph, J.H. Thywissen, R. Younkin, M. Tinkham, M. Prentiss, G.M. Whitesides, *Appl. Phys. Lett.*, **69**, 2773 (1996).

38. C. Gonzalez, S.D. Collins, R.L. Smith, *Transducers*, **1**, 527 (1997).

39. H.Y. Wang, R.S. Foote, S.C. Jacobson, J.H. Schneibel, J.M. Ramsey, *Sensors and Actuators B*, **45**, 199 (1997).

40. H. Lorenz, M. Despont, N. Fahrni, N. LaBianca, P. Renaud, P.J. Vettiger, *Micromech. Microeng.*, **7**, 121 (1997).

41. L. Martynova, L. Locascio, M. Gaitan, G. Kramer, R. Christensen, W. MacCrehan, *Anal. Chem.*, **69**, 4783 (1997).

42. X.M Zhao, Y.N. Xia, G.M. Whitesides, *J. Mater. Chem.*, **7**, 1069 (1997).

43. L. Bousse, A. Kopf-Sill, J.W. Parce, *Transducers*, **97**, 499 (1997).

44. A. Manz, H. Becker, *Transducers*, **97**, 915 (1997).

45. G. Ocvirk, E. Verpoorte, A. Manz, M. Grasserbauer, H.M. Widmer, *Anal. Methods Instrum.*, **2**, 74 (1995).

46. B. He, N. Tait, F. Regnier, *Anal. Chem.*, **70**, 3790 (1998).

47. M. McEnery, J. D. Glennon, J. Alderman, S.C. O'Mathuna, *J. Capill. Electrophor. Microchip Technol.*, **6**, 33 (1999).

48. C. Ericson, J. Holm, T. Ericson, S. Hjerten, *Anal. Chem.*, **72**, 81 (2000).

49. Eksigent, http://www.eksigent.com/hplc/nano.

50. A. Manz, Y. Miyahara, J. Miura, Y. Watanabe, H. Miyagi, K. Sato, *Sensors and Actuators*, **B1**, 249 (1990).

51. S.C. Jacobson, A.W. Moore, J.M. Ramsey, *Anal. Chem.*, **67**, 2059 (1995).

52. A.W. Moore, S.C. Jacobson, J.M. Ramsey, *Anal. Chem.*, **67**, 4184 (1995).

53. F. von Heeren, E. Verpoorte, A. Manz, W. Thormann, *Anal. Chem.*, **68**, 2044 (1996).

54. A. Penrose, P. Myers, K. Bartle, S. McCrossen, *Analyst*, **129**, 704 (2004).

55. H.L. Zeng, H. Li, X. Wang, J.M. Lin, *J. Capill. Electrophor. Microchip Technol.*, **10**, 19 (2007).

56. A. De Mello, *Lab. Chip*, **2**, 48N (2002).

57. A. Manz, D.J. Harrison, E.M.J. Verpoorte, J.C. Fettinger, A. Paulus, H. Ludi, H.M. Widmer, *J. Chromatogr.*, **593**, 253 (1992).

58. D.J. Harrison, A. Manz, Z.H. Fan, H. Ludi, H.M. Widmer, *Anal. Chem.*, **64**, 1926 (1992).

59. D.J. Harrison, Z.H. Fan, K. Seiler, A. Manz, *Anal. Chim. Acta*, **283**, 361 (1993).

60. D.J. Harrison, P.G. Glavina, A. Manz, *Sensors and Actuators B*, **10**, 107 (1993).

61. D.J. Harrison, K. Fluri, K. Seiler, Z.H. Fan, C.S. Effenhauser, A. Manz, *Science*, **261**, 895 (1993).

62. C.S. Effenhauser, A. Manz, H.M. Widmer, *Anal. Chem.*, **65**, 2637 (1993).

63. A.T. Woolley, G.F. Sensabaugh, R.A. Mathies, *Anal. Chem.*, **69**, 2181 (1997).

64. J. Monahan, K.A. Fosser, A.A. Gewirth, R.G. Nuzzo, *Lab. Chip*, **2**, 81 (2002).

65. F.G. Bessoth, O.P. Naji, J.C.T. Eijkel, A. Manz, *J. Anal. Atom. Spectrom.*, **17**, 794 (2002).

66. C.X. Zhang, A. Manz, *Anal. Chem.*, **75**, 5759 (2003).

67. C.J. Ackhouse, A Gajdal, L.M. Pilarski, H.J. Crabtree, *Electrophoresis*, **24**, 1777 (2003).

68. E.B. Cummings, A.K. Singh, *Anal. Chem.*, **75**, 4724 (2003).

69. D. Kaniansky, M. Masar, M. Dankova, R. Bodor, R. Rakocyova, *J. Chromatogr. A*, **1051**, 33 (2004).

70. A. Griebel, S. Rund, F. Schonfeld, W. Dorner, R. Konrad, S. Hardt, *Lab. Chip*, **4**, 18 (2004).

71. Y.C. Wang, M.H. Choi, J. Han, *Anal. Chem.*, **76**, 4426 (2004).

72. Y. Li, J.S. Buch, F. Rosenberger, D.L. DeVoe, C.S. Lee, *Anal. Chem.*, **76**, 742 (2004).

73. S.W. Tsai, M. Loughran, H. Suzuki, I. Karube, *Electrophoresis*, **25**, 494 (2004).

74. J.J. Tulock, M.A. Shannon, P.W. Bohn, J.V. Sweedler, *Anal. Chem.*, **76**, 6419 (2004).

75. E. Olvecka, D. Kaniansky, B. Pollak, B. Stanislawski, *Electrophoresis*, **25**, 3865 (2004).

76. H. Cui, K. Horiuchi, P. Dutta, C.F. Ivory, *Anal. Chem.*, **77**, 1303 (2005).

77. Y.L. Mourzina, A. Steffen, D. Kalyagin, R. Carius, A. Offenhausser, *Electrophoresis*, **26**, 1849 (2005).

78. H.Q. Huang, F. Xu, Z.P. Dai, B.C. Lin, *Electrophoresis*, **26**, 2254 (2005).

79. R. Lin, D.T. Burke, M.A. Burns, *Anal. Chem.*, **77**, 4338 (2005).

80. H. Nagata, M. Tabuchi, K. Hirano, Y. Baba, *Electrophoresis*, **26**, 2247 (2005).

81. H. Cui, K. Horiuchi, P. Dutta, C.F. Ivory, *Anal. Chem.*, **77**, 7878 (2005).

82. G.T. Roman, K. McDaniel, C.T. Culbertson, *Analyst.*, **131**, 194 (2006).

83. D. Kohlheyer, G.A.J. Besselink, S. Schlautmann, R.B.M. Schasfoort, *Lab. Chip*, **6**, 374 (2006).

84. X.Y. Cong, P.C. Hauser, *Electrophoresis*, **27**, 4375 (2006).

85. P.S. Waggoner, H.G. Craighead, *Lab. Chip*, **7**, 1238 (2007).

86. N. Pamme, *Lab. Chip*, **7**, 1644 (2007).

87. A. von Brocke, G. Nicholson, E. Bayer, *Electrophoresis*, **22**, 1251 (2001).

88. A.D. Zamfir, *J. Chromatogr. A*, **1159**, 2 (2007).

89. I. Nischang, U. Tallarek, *Electrophoresis*, **28**, 611 (2007).

90. T.B. Stachowiak, F. Svec, J.M.J. Fréchet, *J. Chromatogr. A*, **1044**, 97 (2004).

91. W.C. Sung, H. Makamba, S.H. Chen, *Electrophoresis*, **26**, 1783 (2005).

92. V.R. Meyer (Ed.), *Practical high performance liquid chromatography*, New York: John Wiley & Sons (1993).

93. G. Guetens, K. van Cauwenberghe, G. DeBoeck, R. Maes, U.R. Tjaden, J. van der Greef, M. Highley, A.T. van Oosterom, E.A. de Bruijn, *J. Chromatogr. B. Biomed. Sci. Appl.*, **28**, 139 (2000).

CHAPTER 2

FABRICATION OF MICROCHIPS

2.1. INTRODUCTION

During the last few decades, information and microfabricated technologies have grown rapidly due to the development of semiconductor integrated circuits (IC). This innovation was explored by Bardeen, Brattain, and Shockley, who shared a Nobel Prize in Physics in 1956. Later on Jack Kilby at Texas Instruments and Robert Noyce at Fairchild Semiconductor independently invented the first integrated circuits. Kilby was awarded the Nobel Prize in Physics in 2000. The idea of integration astonished the whole world and the predication of Moore's law has achieved a great potential in the semiconductor industry for better, more inexpensive, and smaller chips. Later on, the concept of integration became a success in the miniaturization and integration of electronic, mechanical, and microfluidic devices, which led to the fields of microelectromechanical systems (MEMS). Furthermore, this system has achieved a great reputation in microchemical and biomedical (Bio-MEMS) applications, with fast growth within the MEMS market, that is, drug discovery and delivery, biosensors, biochips, lab-on-a-chip systems, nanoparticles, DNA testing and diagnostics, biotelemetry, genomics, and microfluidic devices (micrototal analysis systems [μ-TAS]).

Lab-on-a-chip is the miniaturization and integration of the complete set of devices used in separation science. It most important functions involve sample preparation, reactions, separations, and detection on a single chip. This arrangement was called μ-TAS by Manz in 1990 [1]. We call it nanoanalyses as the

Nanochromatography and Nanocapillary Electrophoresis. By Ali, Aboul-Enein, and Gupta
Copyright © 2009 John Wiley & Sons, Inc.

nature of the separations is at nano and low ranges. The advantages of integration on microchips are reduced size, low power, sample volume, reagent consumption, and manufacture. Moreover, integrations are of better performance in terms of speed, throughput, mass sensitivity, and automation. Besides, these devices are highly effective for poor samples or matrices containing low amounts of ingredients. The movement of fluid in microfluidic devices is controlled by means of micropumps, microvalves, or electroosmosis. For this either microcomponents or electrodes have to be integrated into the flow system. Amer and Badawy [2] discussed the use of MEMS for fabrication of smaller devices that were manufactured by using standard microfabrication techniques (similar to the ones that are used to create computer silicon chips). Many MEMS devices such as microreservoirs, micropumps, cantilevers, rotors, channels, valves, sensors, and other structures have been designed, fabricated, and tested.

Chun et al. [3] discussed fabrication and validation of a multichannel-type microfluidic chip for electrokinetic devices. Silicon glass and polydimethylsiloxane (PDMS) glass microfluidic chips were developed with the unique features of a multichannel. A proper methodology was developed accompanying the deep reactive ion etching as well as the anodic bonding, and optimum process conditions necessary for hard and soft micromachining were presented. Experimentally and theoretically it was shown that a silicon-based microchannel increased streaming potential and higher external current compared to a PDMS-based one. Khan [4] reviewed the applications of laser-based techniques for the fabrication of microfluidic devices for biochips and addressed challenges associated with the manufacturing of these devices. Special emphasis was given to the use of lasers for the rapid prototyping and production of biochips. Efforts were also made on applications on ablation using femtosecond lasers, infrared lasers, laser-induced micro-joining, and the laser-assisted generation of microreplication tools. Besides, microchips are the best pivots among various interdisciplinary areas such as chemistry, physics, material science, biomedicals, and computer and other engineering [5]. Briefly, integration of separation units is an astonishing innovation in separation science. In this chapter we describe the fabrication of microchips used for analyses in chromatography and capillary electrophoresis.

2.2. SUBSTRATES

Certainly, the selectivities, efficiencies, reproducibilities, and applications of nanoliquid chromatography (NLC) and nanocapillary electrophoresis (NCE) machines depend on the materials used for microchips. The microfabrication technologies originated from the microelectronics industry using silicon wafers. Silicon substrates are useful for optical devices, flat panel displays, and semiconducting units. Later on, quartz and glass were employed for fabrication of the microchips. These materials include simple sol-gel, C_2, C_4 and C_{18} and phenyl quartzes. Of course, these materials have a good compatibility with fabrication processes but these have become less popular in due course, being labor intensive and expensive in nature. Sometimes, bonding of the upper and lower parts of the chips is troublesome in these materials. On the other hand, polymeric material such as PDMS is a good material for chips as it contains good biocompatibility, facile bonding ability, high transparency for UV and fluorescence detection, and is cost effective for production. Additionally, it is much less fragile compared to quartz or glass, and can be constructed easily by molding or embossing. In addition, PDMS is a popular material for microfluidic devices due to its surface energy, inexpensiveness, robust processing parameters, gas permeability, biocompatibility, and elastomeric siloxane backbone [6,7]. The high surface energy of PDMS makes its bonding possible with a wide variety of different surfaces by conformal contact between PDMS and the surface. It can also form a reversible conformal bond with glass, metal, or photoresist, which is strong enough to confine fluid within a microfluidic manifold without leaking. For pressure-driven flow plasma or UV oxidation, the surface can be used to bond PDMS irreversibly to a variety of substrates. This sealing process is leak free for moderate liquid flow at high pressures [8,9]. The elastomeric nature of PDMS also enables it to deform around an integrated transducer that may protrude several microns from the planar surface of the substrate, resulting in a reduction of fabrication steps. A disadvantage of PDMS is that many organic molecules and biomolecules are easily adsorbed onto its surface due to the hydrophobic nature of the material. Besides, its hydrophobicity limits the use of many organic solvents in the buffer, except alcohols. Therefore, other materials have been explored and used for fabrication of microchips; these include poly(methylmethacrylate) (PMMA), polycarbonate (PC), polyethyleneteraphthalate (PET), polystyrene (PS), polypropylene (PP), luoroethylene, polyimide (PI), poly(trifluoroethylene) (PTFE), polycyclic olefin copolymer, polyvinyl chloride, fused silica, calcium alginate. Some other auxiliary materials used in chip fabrication are hydrogel, liquid teflon, thermoset polyester, SU-8, and parylene. Common sacrificial materials are photoresist, polyimide, metals, phosphosilicate glass (PSG), and polysilicon. The different polymers provide a wide range of chemical and physical properties, for example, chemical resistance, thermal conductivity, hardness, and dielectric strength to be utilized in nanoanalyses.

The plastic chips are more effective and attractive due to their inexpensiveness with rapid mass productivities. SU-8 has versatile applications in the fabrication of compliant microcomponents due to its outstanding aspect ratio and attainable film thickness, which enables good design of structures such as beams and hinges [10]. Additionally, plastic chips may be disposable, eliminating cross contamination and sample carryover problems in multiple use devices. Soper and coworkers [11–16] and Locascio and coworkers [17,18] have developed plastic microfabricated chips for bioanalytical applications. Parylene is an excellent substance suitable for nano applications due its ease of integration with other microfabrication techniques. It is quite capable of producing microchannels, micropumps, microvalves, filters, pressure and flow sensors, mass flow controllers, electrospray nozzles, and chromatographic and capillary electrophoretic channels [19–24].

2.3. TECHNIQUES OF FABRICATION

The microchip fabrication process involves many steps, which begin with the different patterns, which are projected repeatedly onto the wafer. The major trends and principles of microfabrication are laser technology, lithography, machining, and finishing. These processes are carried out through material selection, wafer fabrication, cleaning, etching, patterning, ion implantation, and packaging. Generally, microfabricated devices are not free standing but usually formed over or in a thicker support substrate. Microchips are patterned by photolithography to form openings and these features are at the micrometer or nanometer scale. The etching is carried out to remove some portion of the thin film or substrate by exposing it to some etching agent such as acids or plasma. The wafer cleaning (surface preparation) of microchips is carried out by thermal diffusion or ion implantation or chemical or mechanical planarization. In microfluidic devices two microchips are fixed together to form microchannels. It is important to mention here that the fabrication processes vary slightly depending on the substrates and the applications.

As discussed above, the fabrication of microfluidic devices is borrowed from microelectronics (micromachining). The manufacturing of these devices started in 1975 in the silicon microprocessors industry with the fabrication of a microchip on a single silicon wafer at Stanford University; it was used in the gas chromatograph [25,26]. Later on the same group [27] miniaturized analytical devices but their applications in separation were not recognized until the 1990s [28]. The standard methods for fabrication of microfluidic devices are described in some textbooks [29–31] and reviews [4,32–43]. The most popular techniques of fabrication for microfluidic chips are photolithography (patterning in photoresist by radiation) [44], embossing/injection or casting molding [45], and complementary metal oxide semiconductor (CMOS), for example, micromachining using isotropic wet etching, reactive ion etching (RIE), and other chemical vapor deposition techniques [46]. About 90% of microfluidic chips reported so far have been fabricated by these techniques. Laser ablation and multi-photon are the latest methods developed for monolithic three-dimensional features, which have provided new ways of integration having many advantages. In addition to this, oxidation, diffusion, ion implantation, chemical vapor deposition (CVD), evaporation, sputtering, wet chemical etching, and dry plasma etching [47] have also been applied to fabricate microfluidics. Some other fabrication techniques have also been developed specifically for MEMS, which include KOH, DRIE (deep reactive-ion-etching), x-ray lithography, electroforming, wafer bonding, electroplating, and 3-D stereo lithography [29]. Büttgenbach et al. [48] described micromachining starting with photolithography as shown in Fig. 2.1, which was used to transfer copies of a master pattern onto the surface of a solid material, such as a silicon substrate, coated with photoresist. The resist-coated substrate is exposed to UV through a mask made of optical quartz glass and coated with a chrome absorber pattern for generating the desired pattern. During this process the photoresist was selectively dissolved. This process transformed the latent resist image formed during exposure into a positive or negative relief image, which served as a masking layer for the etching, thin film deposition, or doping processes. In this way 3-D microstructures can be generated and, finally, the photoresist was completely removed by wet or dry etching.

Recently, polymer MEMS has become popular, resulting in many new techniques suitable for soft lithography [49–51]. Different basic fabrication

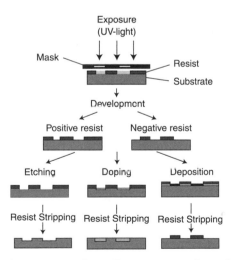

Figure 2.1 Schematic representation of pattern transfer using photolithography [48].

techniques are combined to make a complete device by bulk micromachining [52] and surface micromachining [53]. The former method uses chemical or plasma selective etching with the help of HNA (hydrofluoric acid, nitric acid, and acetic acid), XeF_2, anisotropics such as KOH, TMAH (tetramethylammonium hydroxide), and DRIE. In plasma selective etching microdevices are prepared with the help of sacrificial materials to form free standing or even completely released thin film microstructures, such as microchannels and microcantilevers. Embossing and injection molding methods have also been used to integrate optical components and transducers to microfluidic manifolds [54]. These techniques can mold thermoplastics around embedded features, including emitter tips and fiber optics. In these methods, masters made of silicon or metal are used as a template to mold micromachined devices via pressing the template or heating it against a thermoplastic. The casting is also a useful method of microfluidic fabrication and sensor integration, which provides a route to fabricate microfluidic manifolds with low cost and no mechanical force in comparison to embossing and injection molding. Complementary metal oxide semiconductor (CMOS) processing has also been used to develop microfluidic chips with integrated transducers in both glass and silicon materials. This offers the possibility of integrating multilevel microfluidic channels with a variety of sophisticated electronics and optical features. Man et al. [55] reported the fabrication of a microfluidic device with an integrated CMOS circuit containing control, detection, and communication electronics. The sacrificial etching was used to generate microfluidic channels without requiring a thermal or pressure bonding procedure, which facilitated the micromachined alignment of both optical transducers and fluidic channels. 3-D lithography was suitable for integration of a variety of optical components. Schmidt et al. [56] reported two photon 3-D lithographic techniques for fabrication of waveguides over a printed circuit board, which had a laser and photodiode. This method allowed high precision registration of waveguides with lasers and photodiodes. Many papers are reported in the literature describing the fabrication of microchips for various purposes, which cannot be covered in this chapter. However, some important examples of fabrication on different substrates are discussed in the following sections.

2.3.1. Glass Chips

As discussed earlier, glass was the material of choice for fabrication of microchips during the last decade. Some workers used this material for manufacturing microchips for various purposes. Rodriguez et al. [57] reported the fabrication of a glass microchip device by using standard photolithographic procedures and chemical wet etchings; with sealing of channels via a direct bonding technique. The applications of the device were evaluated by analyzing fluorescein isothiocyanate anti-human IgG. Fang et al. [58] used microscope glass ($20 \times 70 \times 1$ mm) for manufacturing microchips. A 60 mm length of 75 mm i.d., 375 mm o.d. capillary with outer coating was graved as the separation channel. Both ends were inserted 1 mm beyond the walls of two 12 mm sections of 1.5 mm i.d., 2.5 mm o.d. tygon tubing, through holes punctured with a hypodermic stainless steel needle. A 12 mm length of 0.5 mm i.d., 1.6 mm o.d. MicroLine tubing with the capillary tip served as sample/carrier inlet. The tube wall of the downstream tip of the MicroLine tube was carved under a dissecting microscope to produce a conical outlet. The lower end of the section was connected to a short length of 0.6 mm o.d. platinum capillary, which functioned as an electrode, the other end of which was connected to another 5 mm section of $0.5 \mu m$ i.d. MicroLine tube. The chip with polymer covering was then removed from the reservoir using a knife to cut the borders free. The elastomer protruding outside the glass base was then cut off, and the glass rod plugs were removed. Zhang et al. [59] used standard photolithographic and wet chemical etching methods for preparation of a separation channel on glass wafers. The design integrated with sample inlet ports, separation channel, a liquid junction and a guiding channel for the insertion of the electrospray capillary in MS unit. The performance of the device was tested for peptides, proteins, and protein tryptic digests. Tsai et al. [60] described a plasma polymerization technique for modification of a glass chip in nanocapillary isoelectric focusing. The electrophoresis separation channel was machined in Tempax glass chips with length 70 mm, $300 \mu m$ width, and $100 \mu m$ depth. Acetonitrile and hexamethyldisiloxane monomers were used for plasma polymerization (100 nm thickness).

Pu et al. [61] compared the performance of NCE of powder-blasted and hydrogen fluoride-etched microchannels in glass by using rhodamine B and fluorescein as model compounds. The effect of electrical field strength and detection length on the separation efficiency was monitored. The powder-blasted microchannel chips performed well for many applications, although they had low separation efficiency compared to HF-etched chips. Lee et al. [62] reported glass substrates to fabricate the active micro-mixer. The internal residual stress was removed by annealing at 400°C for 1 h. Figure 2.2 represents a schematic illustration of the simplified fabrication process. A thin layer of positive photoresist was used as a mask in the wet chemical etching followed by the lithography patterned and etched in a buffered oxide etchant (BOE). Lee et al. [63] used microscopic glass slides for microfabrication. Schematic representation of the simplified fabrication process is shown in Fig. 2.3. Initially, a thin layer of positive photoresist was used as the etching mask in a wet chemical etching of the glass substrates. The etching mask

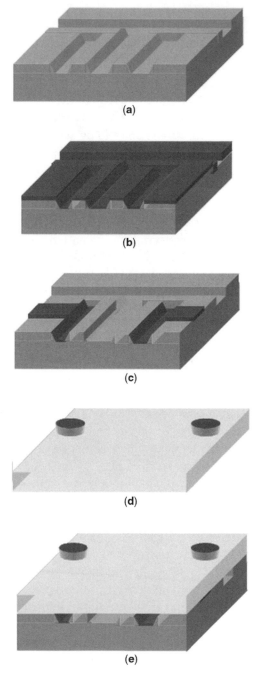

Figure 2.2 A simple fabrication process of active electrokinetically driven micromixers. (a) BOE etching, (b) electron beam evaporation of gold/chromium, (c) gold/chromium etching, (d) cover drilling, and (e) alignment and bonding [62].

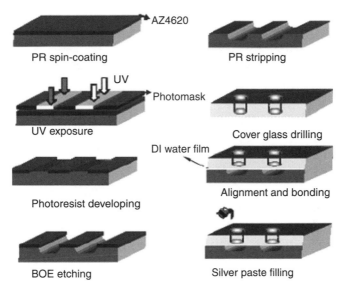

Figure 2.3   Schematic representation of fabrication process for a glass-based microchannel [63].

was applied by using a spin coating process rather than the vacuum deposition commonly used for metal or silicon nitride layers, to reduce the cost and fabrication time. The patterned on substrates was achieved by lithography followed by etching in a 6 : 1 buffered oxide. The holes were drilled in bare glass slides, and then cleaned in a boiling piranha solution also known as piranha etch (mixture of sulfuric acid and hydrogen peroxide). Two glass flats were then carefully aligned and stuck to each other using deionized water, and two plates were thermally bonded in a sintering oven at 580°C for 10 min. Finally, silver conductive material was injected into the microchannels to work as electrodes to establish the field effect within a capacitor.

Brivio et al. [64] described the fabrication of glass microchips whereby the channels were isotropically etched in one or both glass wafers with an HF solution and a chromium-gold mask followed by holes for fluidic connections to the channels blasted in the top wafer. Later on, the processed wafer pair was joined together by fusion bonding. The sizes of the channels were 100 mm wide by 50 mm deep (Fig. 2.4a) and 50 mm wide by 20 mm deep (Fig. 2.4b). Zhang et al. [65] described the fabrication on soda lime glass by using photolithographic and wet chemical etching procedures (Fig. 2.5). The channels were etched to a depth of 20 mm and a width of 60 mm with holes drilled into the etched plate with a 1.2 mm diameter diamond-tipped drill bit at the terminals of the channels. The 4 mm inner diameter and 6 mm tall

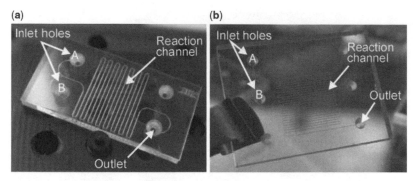

Figure 2.4 Photographs of the glass microreactors: (a) larger (100 mm wide ×
50 mm deep) channels and (b) smaller (50 mm wide × 20 mm deep) channels [64].

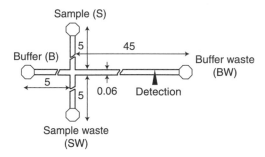

Figure 2.5 Schematic diagram of cross channel microfluidic (dimension in mm) [65].

micropipette tips were epoxied on the chips having the holes and serving as
reservoirs with volumes of approximately 150 mL each. The channels between
the sample reservoir and sample waste reservoir were used for sampling and
the channels between the buffer reservoir and buffer waste reservoir were
used for separation.

Chen et al. [66] described photolithographic and wet chemical etching
methods for fabricating channels onto a 20 × 20 mm glass plate with chromium and photoresist coating. The different types of fabrication processes
are shown in Fig. 2.6. The channels were etched into the plate with dilute
HF/NH_4F within 15 minutes. The channels (Fig. 2.6a to c) for sample solution and organic solvents (Fig. 2.6b and c) were 5 mm long, 25 mm deep,
and 150 mm wide. The extraction channel (Fig. 2.6c and d) was 10 mm
long, 25 mm deep, and 250 mm wide. The access holes for the reservoirs
were drilled into the etched plate with a 1.2 mm diameter diamond-tipped
drill bit at points a and b. The blank glass plates without chromium and

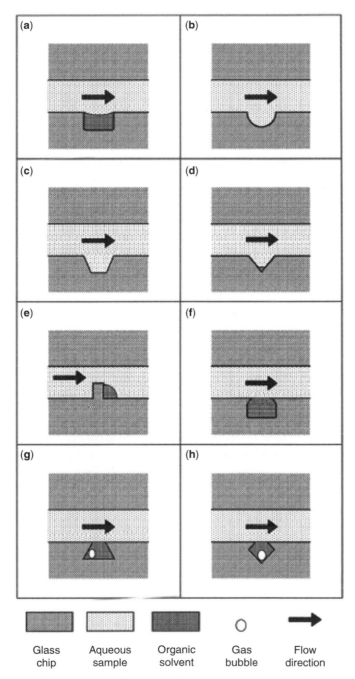

Figure 2.6   Different types of recesses fabricated in the channel wall of the chip designed for trapping organic droplets [66].

photoresist coating were used as cover plates with permanent bonding with epoxy resins. A recess structure with a dimension of 150 mm long, 100 mm wide, and 25 mm deep, with a shrunken opening of 50 mm width, was used. Zhang and Yin [67] designed and fabricated a multi-T microchip on soda lime glass by using photolithographic and wet chemical etching procedures. The channels were etched to a depth of 20 μm and a width of 60 μm and holes were drilled into the etched plate with a 1.2 mm diameter diamond-tipped drill bit at the terminals of the channels. After permanent bonding by a thermal bonding procedure, six 4 mm inner diameter and 6 mm tall micropipette tips were epoxied on the chip surrounding the holes, serving as reservoirs with volumes of approximately 150 μL. The channels between two sample reservoirs and two sample waste reservoirs were used for large volume sampling. The channels between buffer and buffer waste reservoirs were used for separation. Crain et al. [68] designed and developed a glass capillary electrophoresis microchip with integrated electrodes. The processing method described photolithographic patterning and etching using wet chemical processing techniques. The bottom substrate contained seven electrodes required for CE/ECD operation, whereas the top substrate had the microchannel network.

2.3.2. Quartz Chips

Quartz has also been tested for fabrication of microchips. The pros and cons of this material have already been discussed. Ujiie et al. [69] designed and fabricated quartz chips for NCE. The fabrication is shown in Fig. 3.13 of Chapter 3. Chromium films sputtered on quartz 90.5 mm thick and 20 × 20 mm size were patterned by photolithography and wet etching following etching of the capillaries by fluorine-based induction-coupled plasma (ICP). An electrode was biased with a 13.56 MHz power of 15 W to induce sufficient self bias on the quartz surface. After removal of chemicals and chromium, the quartz chip was dipped in 1% hydrofluoric acid and bonded with another quartz plate with four reservoir holes opened by a supersonic machine [70]. Cross-sectional SEM photographs of the etched quartz trench and its expanded view are shown in Fig. 2.7. SEM images of quartz trench patterns etched with 85% C_4H_8-15% SF_6 plasma and etched capillaries pattern with 20 μm width, 50 μm depth, and aspect ratio of 2.5 are shown in Fig. 2.8. Jindal and Cramer [71] fabricated a quartz chip (3 × 3 × 1/16 inches) by using photolithography and wet etching followed by a photo mask for patterning via CorelDRAW®, which was printed out on a transparency using a high resolution printer. The quartz slides were cleaned with acetone, isopropanol, and deionized water and 250 and 1000 Å thick chromium and gold were

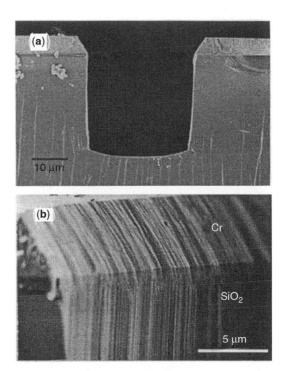

Figure 2.7 SEM photographs of (a) cross-sectional etched quartz trench with 30 μm depth and width and (b) expanded view of the trench sidewall [69].

deposited sequentially on quartz using e-beam evaporation. Gold worked as the hard mask during etching. Later on, hexamethyldisilanzane (HMDS) was spin coated on gold at 3000 rpm for 25 seconds followed with spin coating of the positive photoresist S1813 at 3000 rpm for 40 seconds, with baking at 100°C for 1 min. The channel pattern was transferred photolithographically to the resist via a contact mask aligner. Again the resist was hard baked at 120°C for 30 min. Finally, hydrofluoric and nitric acids were used as quartz etching agents and a depth of 43 μm was obtained in 50 min.

Kitagawa et al. [72] reported the fabrication of IF chips by photolithographic wet etching methods. A polished 0.7 mm thick Pyrex glass or quartz plate was used and after depositions of chromium and gold metal layers, a positive photoresist was spin coated and baked at 90°C for 30 min. Then UV light was exposed through a photomask to transfer the channel pattern onto the photoresist. After etching of the chromium and gold layers, the glass surface was patterned with a 50% HF solution and the remaining photoresist and metals were removed. Another substrate was used as a cover plate and the cover and etched bottom plates were thermally laminated. The

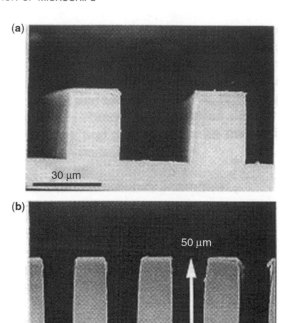

Figure 2.8 SEM micrograph of (a) quartz trench patterns etched with 85% C_4H_8-15% SF_6 plasma and (b) etched capillaries pattern with 20 μm width, 50 μm depth, and aspect ratio of 2.5 [69].

microchip was cut across the microchannel (100 μm width and 40 μm deep) to obtain flat cross sections of the etched channel. To connect a fused silica capillary with dimensions of 50 μm i.d., 365 μm o.d. an access hole (diameter of 400 μm) was drilled. The end faces of the access holes were flattened using a diamond deposited cylindrical high speed tip, for reducing dead volume between the capillary and the microchannel. The microfabrication process on quartz crystal microbalances is shown in Fig. 2.9 [48]. Thin layers of gold were sputter deposited on both surfaces of quartz blanks with the dimension of 1.5 × 1.5 inches with 128 μm thickness. The quartz was chemically thinned down to the desired thickness for high mechanical stability. Hence, circular openings were etched on both sides of the substrate. The quartz was then etched in this area using a mixture of water, hydrofluoric acid, and ammonium fluoride at 60°C. Later on, all layers were stripped before a new gold layer was deposited. Gold electrodes were patterned on both sides of the quartz resonator. In the final step the resonators were cut to their final sizes and affixed to a printed circuit board.

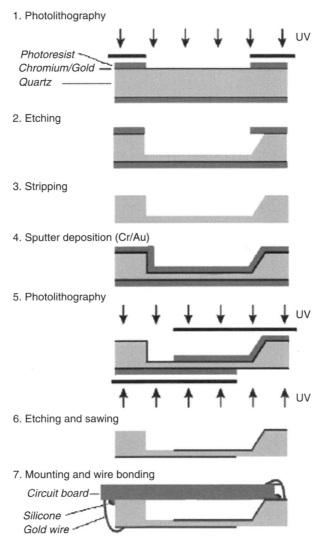

1. Photolithography

Photoresist
Chromium/Gold
Quartz

UV

2. Etching

3. Stripping

4. Sputter deposition (Cr/Au)

5. Photolithography

UV

UV

6. Etching and sawing

7. Mounting and wire bonding

Circuit board

Silicone
Gold wire

Figure 2.9 Schematic representation of fabrication sequence of quartz crystal microbalances [48].

2.3.3. Silica Chips

Silica is a well-known and popular material for fabrication of microchips. The advantages and disadvantages of silica material have already been highlighted in this chapter. Licklider et al. [19] designed and described the electrospray chips shown in Fig. 2.10. A 1.5 μm layer of SiO_2 was grown on both sides of a silica wafer by thermal oxidation. The substrate was patterned using aqueous HF and KOH to form the ports. The top oxide layer was patterned and removed with aqueous hydrofluoric acid. The adhesion of the *p*-xylylene layer was enhanced and the silicon surface was roughened by gaseous BrF_3 and then covered with a thin A-174 silane promotion layer (3-methacryloyloxypropyl trimethoxysilane; SCS). The SCS Model PDS 2010 parylene deposition system was used to coat a 3 μm film of parylene at room temperature at a rate of 3 μm/h, which was patterned with oxygen plasma by using photoresist as the masking material. A 5 μm thick layer of AZ4400 photoresist was used as the sacrificial layer, which was deposited by the standard photoresist spin-coater operation at 2.5 krpm for 40 seconds. After a soft bake at 90°C for 20 min, the lumen structure was defined by standard lithography. Patterned

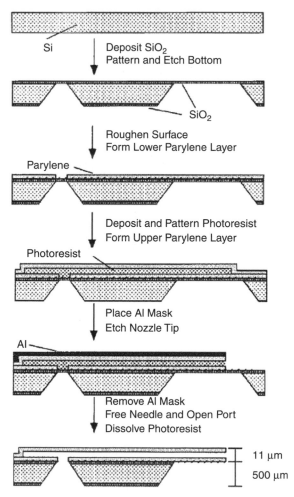

Figure 2.10   Schematic representation of micromachining process used to fabricate microfluidic chips incorporating an electrospray emitter [19].

resist was enclosed by a 3 μm parylene film after treating the first parylene layer with oxygen plasma to promote adhesion between the two parylene layers. This film and aluminum mask was added to protect parylene structures during the plasma etching process, to create the needle tip. The needle tips were defined by oxygen plasma etching, and the wafer was diced. The needle structures were freed by etching with gaseous BrF_3 for removing the underlying silica layer. Finally, the masking layer was removed and the sacrificial photoresist layer was dissolved by acetone for 4 to 6 h. The acetone in the bath was replaced with ethanol and finally with deionized water.

Xie et al. [20] reported the fabrication chip for pumps and an electrospray nozzle. The process used to fabricate the electrochemical pump chips with electrospray nozzle is shown in Fig. 2.11. A 1.5 μm layer of SiO_2 was grown on the surface of a 4 inch silicon wafer by thermal oxidation. The front side oxide layer was patterned and removed with buffered HF. XeF_2 gaseous etching was used to roughen the silicon surface in order to promote the adhesion between subsequent layers and the substrate. The first 4.5 μm parylene layer was deposited.

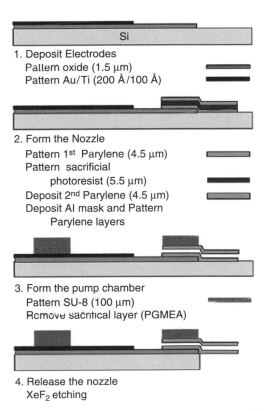

Figure 2.11 Schematic representation of fabrication process for electrochemical pumps on a silicon substrate [20].

A 5.5 μm photoresist layer was patterned as the sacrificial layer, followed by the deposition of a second 4.5 μm parylene layer. The parylene/photoresist/parylene sandwich structure formed the electrospray nozzle and channel when the photoresist was subsequently dissolved. A 1500 Å sputtered aluminum layer was used as a mask for parylene etching to define the shape of the nozzle. Aluminum was removed by a wet etching process. After SU-8 developing, wafers were left inside the SU-8 developer for 2 days to release the photoresist. A serpentine channel (250 μm × 500 μm × 15 mm) extending from the junction of pump channels to the edge of the chip was patterned in the SU-8 layer. Platinum/titanium lines spaced 200 μm apart were patterned under the channel after the electrode deposition step.

2.3.4. Polymer Chips

In addition to those described above, some new polymeric materials have been developed and used in the manufacturing of microchips. These new materials address the drawbacks associated with the above-mentioned materials. Additionally, the advantages of polymeric materials, discussed earlier, compelled scientists to use them as chip materials. McDonald and Whitesides [6] described soft lithography for fabrication of microdevices on PDMS. Figure 2.12 indicates a scheme for rapid prototyping of high resolution transparency as a photomask for generation of the master by photolithography. A system of channels was designed in a CAD program and a high resolution transparency produced, which was used as a photomask in contact photolithography to produce a master, which consisted of a positive relief of photoresist on a silicon wafer and served as a mold for PDMS. This figure also shows liquid PDMS pre-polymer poured over the master and cured for 1 h at 70°C with the PDMS replica peeled from the master and sealed to a flat surface to enclose the channels.

Fu et al. [73] used a hybrid device for fabrication and packaging of microchips. This method consisted of a layer of inorganic substrate like silicon wafer and organic elastomer such as PDMS to perform multiple functions on an integrated chip.

Lagally et al. [74] developed an integrated device for PCR and NCE with electric field control and fluorescence detection. Furthermore, the device had integrated heaters and temperature sensors, as well as PDMS membrane valving to control analyte transport. The PDMS unit was fabricated as membrane valves between a glass PCR chamber and the glass NCE channel, allowing precise control of both process unit operation and analyte transport.

Zhao et al. [75] described a protocol for creating micron sized structures in the substrate material to design, produce, and fabricate a microfluidic system with

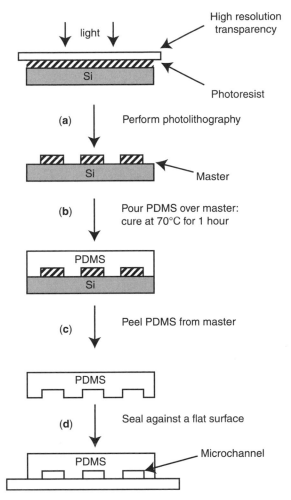

Figure 2.12 Schematic representation of rapid prototyping. A system of channels is designed in a CAD program. (a) Transparency used as a photomask in contact photo-lithography to produce a master, (b) liquid PDMS pre-polymer poured over the master and cured for 1 h at 70°C, (c) PDMS replica peeled from the master, and (d) the replica sealed to a flat surface to enclose the channels [6].

channel features >10 μm in PDMS. The procedure involved the creation of a master template with negative features, using high precision machining. The performance of the device was evaluated for the analysis of DNA fragments.

Zhu et al. [76] designed and fabricated microfluidic devices on polymethyl-methacrylate (PMMA) substrates for electrochemical analysis applications using an improved UV-LIGA process. The microchannel structures were transferred from a nickel mold onto the plastic plates by the hot embossing

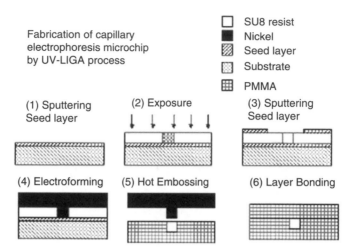

Figure 2.13   Schematic diagram of NCE microchip fabrication by improved ULIGA process [76].

method. During the mold fabrication, the exposure process was optimized for a large ratio of exposed area to unexposed area of negative photoresist (SU-8) followed by nonplanar electroforming technique used for large line space of SU-8 photoresist mold. The electrodes for electrochemical detection were fabricated on other blank PMMA plates through a lift-off process. These substrates with microchannels were bonded to PMMA plates with micro-electrodes by thermal bonding method. Figure 2.13 represents schematically the fabrication of NCE chips by an improved UV-LIGA process.

Nie et al. [77] described designing microchips using CAD 3-D Excalibur software. Initial designs incorporating multiple parallel columns within a single chip were made with a single, double or triple column based electro-osmotic micropump (EOP). By using CAD, designing channels within the PMMA plates were fabricated through direct micromilling with $10 \times 0.4 \times 0.4$ mm channel holding monolithic capillary columns. Two access holes (2×2 mm) were drilled at both ends of the channels for reservoir and outer power connections.

2.3.5. Plastic Chips

Plastic materials have gained importance in microfabrication due to their ease of molding, inexpensiveness, and disposability. Some workers have used these substrates for fabrication of microchips. Pethig et al. [78] and Roberts et al. [79] used laser ablation as a direct method for creating microchannels in plastic chips without the need for fabrication. The methods used an UV excimer laser to burn the microchannels onto the polymer substrate, moving in a predefined, computer-controlled pattern. The channels were sealed with a low film lamination technique. The substrates used in this technology were polystyrene (PS), polyethylene terephthalate (PET), etc. The utility of this method lies in the fact that it has negligible effects on the surface properties of the substrates [16]. By this method high performance sequencing and genotyping devices at relatively low cost could be prepared.

Rossier et al. [80] used UV excimer laser photoablation for changing the surface properties of the plastics and drilling. The authors discussed the method for patterning biomolecules on a polymer along with surface coverage of active antibodies and equilibration time. Besides, a method of designing NCE comprising an on-chip injector, column and electrochemical detector was also discussed. Furthermore, the potential of this disposable device was discussed and compared to classical systems.

Nikcevic et al. [81] reported the fabrication of a microchip by using the UV–LIGA process. The mold for microchip replication was fabricated on a nickel mold disk (7.5 cm. diameter, 1.6 mm thick). The Ni disk was lapped flat and thick photoresist (SU-8 2075 negative photoresist) processing was carried out on the disk followed by Ni electroplating. A plating height of about 100 μm was achieved after 10 h of electroplating in a Ni-sulfamate bath with about 10 μm/h plating rate. After removing the photoresist, the electroplated pattern was polished, which resulted in great improvement in transparency and good perfection in the chips. Replicate plastic microchips were made by injection molding (19,000 psi pressure). The cleaning was achieved by isopropanol and distilled water to remove particulate and organic matter.

Koesdjojo et al. [82] described the fabrication of a microfluidic chip for use in NCE by using a two-stage embossing technique and solvent welding on polymethyl methacrylate with water as a sacrificial layer. The hot embossing method involved a two-stage process to create the final microchip design. Two polymer substrates with different glass transition temperatures (Tg), polyetherimide (PEI) and polymethyl methacrylate (PMMA), were used to make the reusable secondary master and the final chips.

2.3.6. Chips and the Polymerase Chain Reaction

Among various miniaturized analytical devices, polymerase chain reaction (PCR) microchip/microdevices are studied extensively with respect to fabrication, bonding, and sealing. The fabricated PCR microfluidics are very important molecular biological tools to study replication and creation of copies of DNA by cycling through three temperature steps. Normal PCR systems require 1 to 2 h for this purpose because of the low ramping rate resulting from the high thermal capacity of the material in the PCR reaction system, which cannot meet the need for fast DNA amplification in spite of good ramping rates in instruments available in the market nowadays. But, fortunately, chip-based PCR units can replicate the required DNA strand within seconds, which is a landmark in genomic and proteomic research [83]. PCR chip devices are manufactured by silicon micromachining technology, including photolithography, thermal growth of silicon oxide, chemical etching, electrochemical etching, ion etching, chemical vapor deposition (CVD), physical vapor deposition, epitaxy, and anodic bonding. Shin et al. [84] reported the fabrication of a PDMS microchip for PCR. Of course, PDMS is a good material for microdevices but in PCR it causes several problems, such as bubble formation, sample evaporation, and protein adsorption. The authors attempted to overcome these drawbacks by coating micro-PCR chips with parylene film, which has low permeability to moisture and long-term stability. The thermal compability of the PDMS chip was compared with silicon and

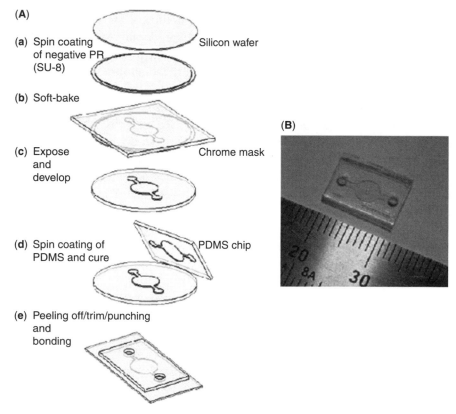

Figure 2.14 Schematic process for (A) the fabrication of a PDMS-based PCR chip and (B) a complete fabricated PCR chip [84].

glass and appropriate thermal responses of PDMS microchips were observed. Figures 2.14A and 2.14B represent schematically the process for fabrication of a PDMS PCR chip and a complete fabricated PCR chip, respectively. Readers interested in the details of PCR microdevices should consult Zhang et al. [38].

2.4. SURFACE MODIFICATION

As in the case of normal chromatography both stationary and mobile phases are also required in NLC. On the other hand, in NCE hydrophilic channel walls with improved control over electroosmotic flow are required for better separation of biological samples. Briefly, the separation efficiencies and selectivities in NLC and NCE depend on the properties of the microchannels, and, therefore, surface modification of the microchannel is usually necessary to achieve good separation of a variety of analytes. Recently, Muck and Svatos [85] reviewed the surface modification of microchannels by critically analyzing the scope and development of chemical modifications based on chemical derivatization or activation of surface layers with reagent solutions, reactive gases, and irradiation. The authors also evaluated bulk modification of polymer chips having incorporation of monomers with selective chemical functionalities throughout the bulk polymer material and integration of the chip modification and fabrication into a single step.

The polymeric chips require special attention to their surface properties due to their poor compatibility with many samples and organic solvents used in forming a coating and in the composition of the mobile phases. Lee et al. [86] and Liu and Chen [87] reviewed the coatings of microchips used in NCE. The authors discussed the advancements and developments of coating preparation methodology and materials, with especial emphasis on the effects of coatings on the resolutions of separation and the reproducibility of separations. Furthermore, dynamic coatings and linked coatings, classified as homopolymers, copolymers, and heterocyclic compounds, were also discussed. The discussion is useful for optimization of separation. The surface modifications of glass microchips are similar to the fused silica capillaries. Normally, coating thickness ranges from microns to angstroms and can be controlled by the amount of dimer. The normal deposition rate of parylene is about 5 μm per hour. The deposition rate is directly proportional to the square of the monomer concentration and inversely proportional to the absolute temperature [19]. High deposition rates can result in poor film quality, which sometimes looks cloudy in contrast to normal clear transparent ones. Shin and coworkers [84] removed the porosity problem of PDMS by coating the chip surface with parylene film. Wang et al. [88] used animation to overcome the aging effects of oxidized PDMS surfaces.

2.4.1. Modification by Polymers

Polymeric materials have been used for surface modification and packing of microchannels in NCE and NLC. Some reports are available on this issue and discussed herein. Hjertén's protocol [89] scems to be the best one for surface control in glass chips by covalently bonded polyacrylamide. Belder et al. [90] reported PVA coating of microchips showing good EOF suppression and reduction in wall analyte interactions. It is important to mention here that these procedures are tedious and time consuming and, hence, have not yet been adopted a standardized protocol for applying these surface modifications in plastic chips. Lin et al. [91] described PMMA coating by using poly(vinylpyrrolidone) (PVP), poly(ethylene oxide) (PEO), and 13 nm gold nanoparticles (GNPs) and tested for DNA separation. Landers' group [92–94] reported dynamic coating of microchannels with adsorptive polymers such as dynamic mechanical analysis (DMA), DEA, etc. Lai et al. [95] used a gas resin injection procedure for modifying channels of PMMA and tested for analysis of DNA. The procedure involved the injection of hydroxyethyl methacrylate monomer (HEMA), a photoinitiator, and other additives (sodium dodecyl sulfate, ethylene glycol, and polyethylene glycol [PEG]) into the microchannel by applying nitrogen pressure. The liquid was forced out leaving behind a thin layer on the microchannel walls. Later on, resin was cured with UV light resulting in more hydrophilic unmodified PMMA surface. A mixture of poly-HEMA and 10% PEG was found to be most hydrophilic coating as determined by contact angle measurements. Barker et al. [96,97] applied polyelectrolyte multilayers composed of PSS and poly(allylamine hydrochloride) for EOF control in PS and PETG microfluidic devices. Liu et al. [98] described dynamic coatings of polyelectrolyte multilayers to overcome some disadvantages of PDMS, such as the tendency of PDMS surfaces to absorb analytes and the instability of electroosmotic flow in oxidized PDMS channels. It was possible with multilayers of cationic polybrene and anionic dextran sulfate in PDMS microchannels and led to stable EOF over a wide pH range. Ro et al. [99] described polyelectrolyte multilayers to improve the performance of PDMS devices for packed bed in NCEC.

Belder et al. [90] described poly(vinyl alcohol) coated microfluidic devices in NCE for a suppressed electroosmotic flow and improved separation performance of labeled amines. A threefold increase in separation efficiencies was obtained on coated chips. In PVA-coated channels, rinsing or etching steps could be omitted, which are necessary for uncoated devices. Wu et al. [100] described multilayer poly(vinyl alcohol)-adsorbed coating on poly(dimethylsiloxane) microfluidic chips for biopolymer separation. The reflection absorption infrared spectrum (RAIRS) showed that 88% hydrolyzed polyvinyl alcohol (PVA) adsorbed more strongly than 100% hydrolyzed PVA on the oxygen plasma pretreated PDMS surface. Repeating the coating procedure three times was found to produce the most robust and effective coating. PVA coating converted the original PDMS surface from a hydrophobic to a hydrophilic surface, and suppressed electroosmotic flow (EOF) in the range of pH 3 to 11.

Pumera et al. [101] used gold nanoparticles to improve the selectivities of solutes and efficiency in NCE. The microchannel wall of a microfluidic device was coated with a layer of poly(diallyldimethylammonium chloride) (PDADMAC) and citrate-stabilized gold nanoparticles were collected on it. The authors reported that the resolutions and the plate numbers of the solutes were doubled in the presence of the gold nanoparticles. Qu et al. [102] described chemical modification of a PMMA microchannel surface by protein patterning on a plastic microfluidic channel. A craft copolymer was designed and synthesized to introduce silane functional groups onto the plastic surface. Thus, anchoring of proteins onto the hydrophobic PMMA microchannels could be realized with bioactivity preserved. The performance of protein patterning in a microfluidic channel was evaluated by performing NCE with laser-induced fluorescence (LIF) detection and confocal fluorescence microscopy, for trypsin digest. Jiang et al. [103] described a method for avidin and biotin immobilized structures. Gunawan et al. [104] fabricated a stable and covalently proteins immobilized microdevices. Wang et al. [105] reported protein modification of poly(dimethylsiloxane) microfluidic channels for the enhanced separation in NCE by coating albumin and lysozyme proteins. The separated compounds were neurotransmitters and the environmental pollutants, which were efficiently separated within 140 seconds in a 3.7 cm long separation channel.

Du and Fang [106] reported a simple and robust static adsorptive (dynamic) coating process using 2% hydroxyethylcellulose for surface modification of PMMA microfluidic chips for DNA separation. The authors evaluated the performance of the coated chips at different phases of the coating process by consecutive gel electrophoretic separations with LIF detection using a PhiX-174/ HaeIII DNA digest sample. The authors reported good performance of the coated chips even after three months. The average precision of migration time was 1.31% RSD ($n = 6$) during a 25 day study. Lee et al. [107] described the surface modification of microchannels in NCE PDMS, quartz, and glass materials. The authors reported that the new functional groups could be formed on the PDMS surface after treatment, resulting in a change in the surface property. The time-dependent surface property of the plasma-treated PDMS was measured in terms of the zeta potential. Two simple and reliable methods utilizing organic-based spin on glass and water-soluble acrylic resin were reported. This method was evaluated by using PhiX-174 DNA maker separation, indicating the surface property modification and separation efficiency improvement. Liu and Henry [108] utilized the chemical modification of the microchannels to reduce or eliminate the analyte wall interactions and alteration of electroosmotic flow. The authors reported a stable polyelectrolyte multilayer coating to prevent analyte adsorption for the rapid and efficient separation of biomolecules within microchannels. The method of microchip coating with polyelectrolytes and the expected results have been discussed. Brown et al. [109] reported different types of surface modifications of PMMA to change substrates properties. The methods include air plasma treatment, acid-catalyzed hydrolysis, and aminolysis. The resulting substrates were carboxyl- and amine-terminal PMMA surfaces.

2.4.2. Modification by Silica Gel

Silica gel is the famous stationary phase in liquid chromatography and capillary electrophoresis. It has been used to coat and pack microchannels in a number of studies. Jacobson et al. [110] coated the wall of a glass microchip (a 16.5 cm long serpentine channel that was 5.6 μm deep and 66 μm) with octadecyltrichlorosilane in a pressurized, high temperature process for C_{18} nature. Later on, an improvement in coating with octadecylsilane modification procedure was reported at room temperature [111] and chips with C_{18} coating exhibited a decrease of 10% to 25% in electroosmotic flow compared to non-coated channels. Constantin et al. [112] reported sol gel chemistry of thin silica gel coatings having high surface areas due to their porous structure. A partially gelled mixture of tetraethyoxysilane (TEOS) and n-octyltriethoxysilane (C_8-TEOS) was injected into the channels of a glass microchip for coating purposes. Silica gel was anchored to the walls of the chip by condensation reactions involving surface silanol groups. The channels were flushed with solvent to remove excess gel, leaving only a thin layer of coating on the wall surface. Sodium hydroxide was used to increase the surface area of the coating. Soper et al. [113] described a two-step procedure for forming octa-decylated surfaces on hot embossed PMMA devices. The surface methyl ester groups of polymer were converted into amines by reaction with N-lithioethylenediamine followed by filling with octadecylisocyanate, which reacted with pendant amine functionalities to afford a C_{18} modified surface. The surface resulted in a positively ionized nature due to residual unreacted surface amine groups resulting in electroosmotic flow from cathode to anode. Due to the large surface area of collocated monolith support structures (COMOSS), silica-based stationary phases are widely characterized and easily functionalized. However, the uniform packing and reproducibly in the channels is a significant technical challenge [114]. He and coworkers [115,116] coated COMOSS columns with silanized (3-aminopropyl)triethoxysilane. Then poly(styrene sulfonate) was attached electrostatically to the amine-modified surface to create a hydrophobic coating suitable for the reverse phase separation mode. Jindal and Cramer [71] fabricated a quartz chip (3 × 3 × 1/16 inches) by using photolithography and wet etching followed by packing with silica gel. Stationary phase particles were immobilized through the sol gel method, prepared by mixing TEOS (20 μL), HCl (0.12 M, 10 μL), and DEME (70 μL). The C_4 silica (250 mg/mL) was added to the above solution and kept for 5 h followed by agitation and packing of the channel. Figure 2.15 represents schematically the process of immobilization of the stationary phase in the channel. It also indicates the microfabricated quartz chip with cross channel design, PDMS slab bonded reversibly to the quartz, and quartz with stationary phase particles immobilized in the separation channel after removing the PDMS cover.

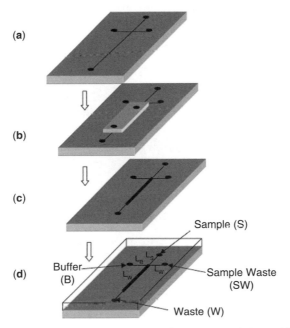

Figure 2.15 Schematic representation for the process of immobilization of the stationary phase in the channel. (a) Microfabricated quartz chip with cross channel design; (b) PDMS slab bonded reversibly to the quartz, defining the location of the stationary phase; (c) quartz with stationary phase particles immobilized in the separation channel after removing the PDMS cover; (d) bonding of PDMS and quartz after oxygen plasma treatment [71].

2.4.3. Modification by Monolithics

Recently, columns made of a single piece of monolithic silica were introduced as an alternative to particle-based columns. These columns possessed a biporous structure consisting of larger macropores (2 μm) that permit high flow rates with low back pressure. It is then possible to perform analysis with high linear flow velocity but without significantly reduced separation efficiency. The monolithic stationary phases have many advantages of high surface area and easily controlled surface chemistry. Besides, the monolith can be prepared easily and rapidly via free radical polymerization within the channels of the microdevice. The porosity, surface area, and pore size of the monolith are controlled by adjusting the composition of the initial monomer solution and the polymerization conditions. Additionally, a wide variety of chemistries can be incorporated into the microfluidic device through suitable selection of monomers. The performance of the monolithic stationary phase in the microchip was compared to the fused silica. These features are compatible for an ideal coating and packing of microchannels. Some authors reported surface modification and packing of channels using these substrates. Ericson et al. [117] described polymer-based monolithic stationary phases packed in quartz microchannels. Svec and colleagues [118,119] reported UV-initiated polymerization for the preparation of monoliths within microfluidic devices. Later this group refined this approach for the preparation of the monolith and in situ surface functionalization [120–124]. The process was carried out with the preparation of a variety of methacrylate-based monolithic porous polymer stationary phases. Breadmore et al. [125,126] reported monolithic stationary phases in NCEC; basically, it is a modification of the procedure described by Ishizuka et al. [127].

2.4.4. Modification by Sulfonation

Some authors modify the surface chemistries of microchips by certain chemical reactions. Among them sulfonation is one of the mostly used methods. Soper et al. [113,128] used sulfur trioxide gas to modify the hydrophobicity of microchannels. Furthermore, the same group [113,129] modified PMMA surface properties by sulfonation via fuming sulfuric acid, resulting in low EOF velocity. In the same way, the authors used o-diaminoalkanes to achieve an amine-terminal PMMA surface. Xu et al. [130] reported that coating of PMMA channels with hydroxypropyl methylcellulose (HPMC) controlled EOF resulting into a good separation of DNA molecule. Vaidya et al. [128] made polycarbonate surfaces more hydrophilic and manipulated electroosmotic flow by chemical treatment with sulfur trioxide gas.

2.5. DESIGNS OF CHIPS

The shape, size, and length of microchannels are important issues to achieve good separation in NLC and NCE. Therefore, different types of microchips with various dimensions of microchannels were designed and fabricated. Normally, an NLC microchip has a wafer access hole on the back as fluidic inlet and solvents flow in following entry into the electrospray nozzle with minimum dead volume. The size of the column varied depending on the requirement. Bousse et al. [131] described a chip-based electrokinetic device consisting of parallel series of separation channels, which are capable of distributing a serial sample into these channels, resulting in good separation of various compounds. Similarly, Manz and Becker [132] reported a serial to parallel converter for NCE used for efficient separations.

He et al. [115] designed and fabricated a special type of column called col-located monolith support structures (COMOSS) which are essentially a tightly packed array of posts (Fig. 2.16). The size, shape, and dimensions of the channels can be varied depending on requirements. COMOSS have many advantages, such as the uniformity and regularity of the support particles, high level of control over the channel dimensions and geometry, and the ability to control the extent of mixing in the column. Besides, COMOSS columns showed two to three times the surface area of normal columns, resulting in higher loading capacities. He and coworkers [115,116] investigated the design and fabrication of the devices. Anisotropic deep reactive ion etching was used to create small, high aspect ratio support structures in quartz devices (150 μm wide and 4.5 cm long). However, it is important to mention that COMOSS devices are costly (about US$ 2000) [125], which may be considered a drawback.

Slentz et al. [133] described the effects of geometry (size, shape, and dimensions) on the performance of COMOSS. Vreeland and Barron [134] described the design of functional materials for genomic and proteomic analyses in NCE. The authors discussed different polymer chemistries for microchannel surface passivation and improved DNA separation.

Novel bioconjugate materials designed specifically for NCE have also been discussed. Sato et al. [135] described a bead bed immunoassay system in NCE. The chip has branching multichannels and four reaction and detection regions. This device was capable of processing four samples at a time. The biases of the signal intensities obtained from each channel were within 10%, and coefficients of variation were almost the same level as the single straight channel assay.

Ceriotti et al. [136] developed a poly(dimethylsiloxane)-based fritless NCE with conventional particulate stationary phases. The device includes an injector and a tapered column with particles of the stationary phase.

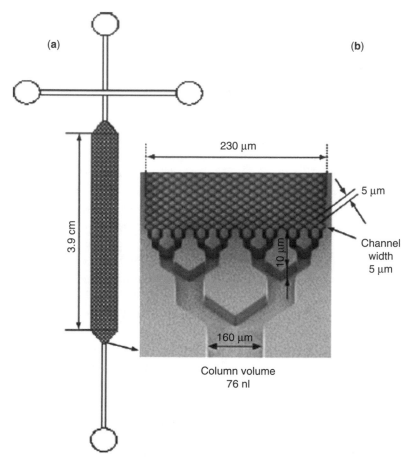

Figure 2.16   Representation of (a) layout of a typical COMOSS separation column and (b) an SEM image showing the microfabricated inlet splitter [115].

The separation channel was packed by 3 μm, octadecylsilanized silica microspheres using a vacuum system. The packing was stabilized in the column by a thermal treatment, and its stability and quality were evaluated by indirect fluorescence detection. The effects of voltage on electroosmotic flow and efficiency were investigated. Swinney and Bornhop [137] described quantification and evaluation of joule heating in NCE. According to the authors a 7% reduction in separation efficiency was determined for a high current drawing buffer such as Tris-boric acid under an applied field of just 400 V/cm. Furthermore, it was realized by the authors that heating effects on the chip NCE were underestimated, and there was a need to re-address the theoretical model. Fu et al. [138] reported the geometry effects on band spreading in NCE. The authors have designed and tested various geometric bend ratios to greatly reduce racetrack effect. The effects of the separation channel geometry, fluid velocity profile, and bend ratio on the band distribution in the detection area were discussed. A folded square U-shaped channel was reported to be better for miniaturization and simplification. The band tilting was corrected and the racetrack effect reduced in the detection area when the bend ratio was 4:1. The detection time obtained from the present numerical solution matched well with the experimental data. According to the authors the geometry and flow field conditions in the separation microchannel of NCE might have an important impact on the system separation efficiency. Chen et al. [139] developed 2-D NCE via assembling, disassembling, and reassembling of PDMS chip components with perpendicular separation channels. Thorsen et al. [140] designed and developed microfluidic networks as in the case of electronic integrated circuits, having 1024 flow channels addressed by only 20 control channels and a multiplexed valve system.

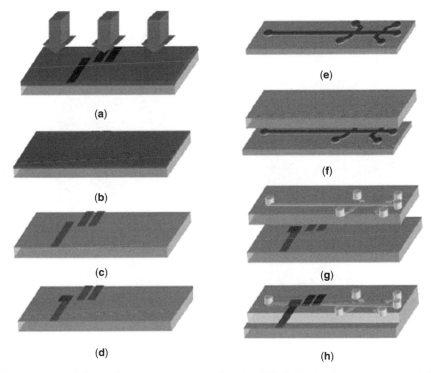

(a)

(b)

(c)

(d)

(e)

(f)

(g)

(h)

Figure 2.17 Schematic representation of a simplified fabrication process used for proposed GNEE NCE microchip. (a) Contact electrode exposure, (b) electrode developing, (c) platinum etching, (d) GNEE direct bonding, (e) master fabrication, (f) hot embossing, (g) alignment, and (h) low temperature solvent bonding [146].

Gitlin et al. [141] designed topographical features on one channel wall to achieve electrokinetic pumping by applying the field across the channel. Liu et al. [142] reported a biochip having mixers, valves, pumps, channels, chambers, heaters, and DNA microarray sensors for effective analysis of DNA strands. Shaikh et al. [143] described a modular microfluidic structure for biochemical analysis. Park and Madou [144] designed 3-D electrode for high throughput dielectrophoretic concentration, filtration, and separation. Taff and Voldman [145] described scalable architecture for trapping, imaging, and sorting individual microparticles by using a positive dielectrophoretic (p-DEP) trapping array. The authors demonstrated capturing, holding, and release operations with both beads and cells in small arrays of this new architecture. Liao et al. [146] described a fabrication process for gold nano-electrode ensemble (GNEE) in NCE

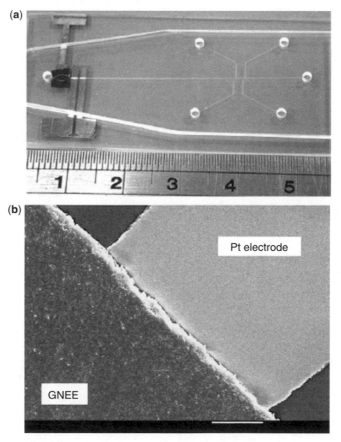

Figure 2.18   Representation of (a) photograph of completed GNEE NCE microchip and (b) SEM image of bonded GNEE on Pt electrode [146].

microchip. Microchannels were created by using a hot embossing process with a glass etched mold. Figure 2.17 shows the simplified fabrication process for a GNEE NCE microchip. Figures 2.18a and b indicate a photograph of the completed GNEE NCE microchip and SEM image of bonded GNEE on a platinum electrode. PCR microdevices are single- or multi-chamber units, which are single-chamber stationary, multi-chamber stationary PCRs. Flow through and thermal convection driven PCR microfluidics and on-chip serpentine rectangular channel base have also been developed. Zhong et al. [147] used a two-side etching/alignment and a simplified one-side/two-step etching method for designing the microfabricated two weir structures within a microchannel. The former method required a straightforward alignment step while the latter technique comprised a simplified wet etching process using paraffin wax as the temporary protective layer. The authors reported good separation of λ-DNA extractions on these chips. Shi et al. [148] described radial microchannels in NCEs for DNA sequencing consisting of 16, 96, and 384 channels (Fig. 8.10, Chapter 8). The authors studied geometry both theoretically and experimentally to reduce turn dispersion and minimize turn-induced resolution loss. The radii of curvature of the turn, the tapering length, and the degree of tapering were optimized.

2.6. BINDINGS IN CHIPS

In NLC and NCE graved structures are required in the form of columns and capillaries. Therefore, the binding of two chips to form microchannels is an important issue. The sealing of the chips must be leakage free and, hence, different approaches have been used for this purpose, including pressurizing, gluing, and mechanical holding. Stjernstrom and Roeraade [149] reported thermal bonding in soda lime glass. Berthold et al. [150] described glass bonding by the deposition of silicon nitride, polysilicon, or amorphous silicon on the glass surfaces. Huang et al. [151] used UV curable glue for bonding glass chips at room temperature and verified its performance in NCE. Ito et al. [152] used a water glass solution for glass binding. Jia et al. [153] reported a room temperature bonding process for the fabrication of glass chips resulted in more than 95% bonding quality. Satyanarayana et al. [154] described UV-curable adhesive bonding for silicon, glass, and nitride surfaces. Howlader et al. [155] reported the bonding of glass wafers by using the surface activated bonding method at room temperature without heating.

Nakanishi et al. [70] pressed two quartz chips at a load of 1.3 MPa at room temperature for 24 h to produce enclosed channels. Satoh [156] used a low temperature and low external load technique for silicon–silicon bonding using water glass. Weinert et al. [157] used oxygen and argon plasma for binding of chips made of silicon, silicon dioxide, and quartz. Wild et al. [158] used laser radiation to bond silicon and glass wafers. Wu et al. [159] transferred a thin layer of adhesive to a patterned substrate for sealing PDMS to a flat plate to form a microchip. Hui et al. [160] used microwave plasma, generated in a glass bottle containing 2 to 3 torr oxygen, for irreversible sealing in PDMS. Yussuf et al. [161] used microwave energy for binding two PMMA chips. Blom and coworkers [162] exploited thermal stress caused by the difference in thermal expansion to join two channels. Kelly and coworkers [163] used high internal pressures to bind polymeric microchips. Pattekar and Kothare [164] reported microfluidic interconnectors in Teflon capillaries by sealing with a high temperature curing epoxy. Lee et al. [165] reported electrochemical discharge for microfluidic connections in plastic chips. Jindal and Cramer [71] joined together two quartz chips ($3 \times 3 \times 1/16$ inches) by drilling access holes with a diamond-tipped drill. Nie et al. [77] drilled two access holes (2×2 mm) at both ends of the channels for reservoir and outer power connections. After cleaning, 10×0.1 mm monolithic silica was carefully inserted into the channels. The gap between channel and column was sealed at either end by using epoxy resin to prevent leakage. Similarly, Nikcevic et al. [81] developed and patterned a wafer (channel width $100 \,\mu$m and depth $80 \,\mu$m) by covering the wafer (IM-PMMA, 0.9 mm) through thermoplastic fusion bonding ($86°$C, 0.2 tons pressure). Reservoirs were made by drilling holes at appropriate locations on the fabricated wafers; for loading analytes and applying operating voltages. To increase the volume of reservoirs, polypropylene pipette tips or glass tubes ($1/4$ in. i.d. $\times 7$ mm height) pieces were bonded to wafer holes with epoxy after chip bonding.

2.7. CONCLUSION

The world of the microchip has changed the lifestyle of society since the discoveries of Bardeen, Brattain, Shockley, Kilby and Noyce in silicon chip fabrication. Various substrates and techniques have been used to fabricate microchannels but polymeric materials are ideal in NLC and NCE applications. Surface modifications are also an important issue for achieving good separations and many materials have been used for these purposes. The developed chips and microchannels are capable of separating a wide variety of compounds at nano or low level ranges. The microfluidic devices are not fully developed and are still in development stages. We anticipate future development of new materials and methodologies that will have an impact on the fabrication of microchips and separation science, including interdisciplinary areas.

REFERENCES

1. A. Manz, N. Graber, and H.M. Widmer, *Sensors & Actuators B: Chem.*, **1**, 244 (1990).
2. S. Amer, W. Badawy, *Curr. Pharm. Biotechnol.*, **6**, 57 (2005).
3. M.S. Chun, M.S. Shim, N.W. Choi, *Lab. Chip*, **6**, 302 (2006).
4. M.C.G. Khan, *Anal. Bioanal. Chem.*, **385**, 1362 (2006).
5. N.P. Mahalik, *Micromanufacturing and nanotechnology*, New York: Springer (2006).
6. J.C. McDonald, G.M. Whitesides, *Acct. Chem. Res.*, **35**, 491 (2002).
7. C.S. Effenhauser, G.J. Bruin, A. Paulus, M. Ehrat, *Anal. Chem.*, **69**, 3451 (1997).
8. L. Chen, J. Ren, R. Bi, D. Chen, *Electrophoresis*, **25**, 914 (2004).
9. R.S. Martin, A.J. Gawron, S.M. Lunte, C.S. Henry, *Anal. Chem.*, **72**, 3196 (2000).
10. V. Seidemann, J. Rabe, M. Feldmann, S. Büttgenbach, *Microsyst. Technol.*, **8**, 348 (2002).
11. S.A. Soper, D.C. Williams, Y.C. Xu, S.J. Lassiter, Y.L. Zhang, S.M. Ford, R.C. Bruch, *Anal. Chem.*, **70**, 4036 (1998).
12. S.A. Soper, S.M. Ford, Y.C. Xu, S. McWhorter, S.J. Lassiter, D. Patterson, R.C. Bruch, *J. Chromatogr. A*, **853**, 107 (1999).
13. C.S. McWhorter, S.A. Soper, *Electrophoresis*, **21**, 1267 (2000).
14. S.J. Lassiter, W. Stryjewski, B.J. Legendre, R. Erdmann, M. Wahl, J. Wurm, R. Peterson, L. Middendorf, S.A. Soper, *Anal. Chem.*, **72**, 5373 (2000).
15. M.B. Wabuyele, S.M. Ford, W. Stryjewski, J. Barrow, S.A. Soper, *Electrophoresis*, **22**, 3939 (2001).
16. Y.C. Xu, B. Vaidya, A.B. Patel, S.M. Ford, R.L. McCarley, S.A. Soper, *Anal. Chem.*, **75**, 2975 (2003).
17. D.L. Pugmire, E.A. Waddell, R. Haasch, M.J. Tarlov, L.E. Locascio, *Anal. Chem.*, **74**, 871 (2002).
18. D. Ross, L.E. Locascio, *Anal. Chem.*, **74**, 2556 (2002).
19. L. Licklider, X.Q. Wang, A. Desai, Y.C. Tai, T.D. Lee, *Anal. Chem.*, **72**, 367 (2000).
20. J. Xie, Y. Miao, J. Shih, Q. He, J. Liu, Y.C. Tai, T.D. Lee, *Anal. Chem.*, **76**, 3756 (2004).
21. X.Q. Wang, Y.C. Tai, *Normally closed in-channel micro check valve, Proceedings of the 13th IEEE International conference on micro electro mechanical systems (MEMS 2000)*, Miyazaki, Japan, 68 (2000).
22. J. Xie, X. Yang, X.Q. Wang, Y.C. Tai, *Surface micromachined leakage proof parylene check valve, Proceedings of the 14th IEEE International Conference on*

Micro Electro Mechanical Systems (MEMS 2001), Interlaken, Switzerland, 539 (2001).

23. J. Xie, J. Shih, Y.C. Tai, *Integrated surface-micromachined mass flow controller, Proceedings of the 16th IEEE International Conference on MicroElectro Mechanical Systems (MEMS 2003)*, Kyoto, Japan, 20 (2003).

24. H.S. Noh, P.J. Hesketh, G.C. Frye-Mason, Parylene gas chromatographic column for rapid thermal cycling, *J. Microelectromechanical Systems*, **11**, 718 (2002).

25. S.C. Terry, Ph.D. Thesis, Stanford University, Stanford, CA (1975).

26. S.C. Terry, J.B. Angell, *A column gas chromatography system on a single wafer of silicon, Theory, Des. Biomed. Appl. Solid State Chem. Sens., Workshop 1978,* 207 (1977).

27. S.C. Terry, J.H. Jerman, J.B. Angell, *IEEE Trans. Electron. Devices*, **26**, 1880 (1979).

28. J.M. Ramsey, *Nat. Biotechnol.*, **17**, 1061 (1999).

29. M.J. Madou, *Fundamentals of microfabrication: The science of miniaturization*, 2nd ed., Boca Raton: CRC Press (2002).

30. P. Rai-Choudhury, *Handbook of microlithography, micromachining, and microfabrication*, 1st ed., Vol. 1, Bellingham, WA: SPIE Press (1997).

31. P. Rai-Choudhury, *Handbook of microlithography, micromachining, and microfabrication*, 1st ed., Vol. 2, Bellingham, WA: SPIE Press (1997).

32. M. Rothschild, M.W. Horn, C.L. Keast, R.R. Kunz, V. Liberman, S.C. Palmateer, S.P. Doran, A.R. Forte, R.B. Goodman, J.H.C. Sedlacek, R.S. Uttaro, D. Corliss, A. Grenville, *Lincoln Lab. J.*, **10**, 19 (1997).

33. D.R. Reyes, D. Iossifidis, P.A. Auroux, A. Manz, *Anal. Chem.*, **74**, 2623 (2002).

34. E.A.S. Doherty, R.J. Meagher, M.N. Albarghouthi, A.E. Barron, *Electrophoresis*, **24**, 34 (2003).

35. C.W. Kan, C.P. Fredlake, E.A.S. Doherty, A.E. Barron, *Electrophoresis*, **25**, 3564 (2004).

36. T. Vilkner, D. Janasek, A. Manz, *Anal. Chem.*, **76**, 3373 (2004).

37. M. Pumera, *Talanta*, **66**, 1048 (2005).

38. C. Zhang, J. Xu, W. Ma, W. Zheng, *Biotech. Adv.*, **24** 243 (2006).

39. P.S. Dittrich, K. Tachikawa, A. Manz, *Anal. Chem.*, **78**, 3887 (2006).

40. P. Abgrall, V. Conedera, H. Camon, A.M. Gue, N.T. Nguyen, *Electrophoresis*, **28**, 4539 (2007).

41. B.A. Peeni, M.L. Lee, A.T. Woolley, *Electrophoresis*, **27** 4888 (2006).

42. T.B. Stachowiak, F. Svec, J.M.J. Frechet, *J. Chromatogr. A*, **1044**, 97 (2004).

43. G.T. Roman, R.T. Kennedy, *J. Chromatogr. A*, **1168**, 170 (2007).

44. D.C. Duffy, J.C. McDonald, J.A. Schueller, G.M. Whitesides, *Anal. Chem.*, **70**, 4974 (1998).

45. Y. Wang, B. Vaidya, H.D. Farquar, W. Stryjewski, R.P. Hammer, R.L. McCarthy, S.A. Soper, *Anal. Chem.*, **75**, 1130 (2003).

46. S.C. Jacobson, J.M. Ramsey, *Anal. Chem.*, **68**, 720 (1997).

47. S. Wolf, R.N. Tauber, *Silicon processing for the VLSI era*, Sunset Beach, CA: Lattice Press (2000).

48. S. Büttgenbach, M. Michalzik, R. Wilke, *Eng. Life Sci.*, **6**, 449 (2006).

49. Y.N. Xia, G.M. Whitesides, *Annu. Rev. Mater. Sci.*, **28**, 153 (1998).

50. G.M. Whitesides, E. Ostuni, S. Takayama, X.Y. Jiang, D.E. Ingber, *Annu. Rev. Biomed. Eng.*, **3**, 335 (2001).

51. J.C. McDonald, D.C. Duffy, J.R. Anderson, D.T. Chiu, H.K. Wu, O.J.A. Schueller, G.M. Whitesides, *Electrophoresis*, **21**, 27 (2000).

52. G.T.A. Kovacs, N.I. Maluf, K.E. Petersen, *Proc. IEEE*, **86**, 1536 (1998).

53. J.M. Bustillo, R.T. Howe, R.S. Muller, *Proc. IEEE*, **86**, 1552 (1998).

54. S. Qi, X. Liu, S. Ford, J. Barrows, G. Thomas, K. Kelly, A. McCandless, K. Lian, J. Goettert, S.A. Soper, *Lab. Chip*, **2**, 88 (2002).

55. P.F. Man, D.K. Jones, C.H. Mastrangelo, *Proc. Int.Workshop on MEMS*, Nagoya, Japan, 311 (1997).

56. V. Schmidt, L. Kuna, V. Satzinger, R. Houbertz, G. Jakopic, G. Leising, *Proc. SPIE*, **6476**, 64760P (2007).

57. I. Rodriguez, Y. Zhang, H.K. Lee, S.F. Li, *J. Chromatogr. A*, **781**, 287 (1997).

58. Q. Fang, F.R. Wang, S.L. Wang, S.S. Liu, S.K. Xu, Z.L. Fang, *Anal. Chim. Acta.*, **390**, 27 (1999).

59. B. Zhang, F. Foret, B.L. Karger, *Anal. Chem.*, **72**, 1015 (2000).

60. S.W. Tsai, M. Loughran, A. Hiratsuka, K. Yano, I. Karube, *Analyst*, **128**, 237 (2003).

61. Q.S. Pu, R. Luttge, H.J. Gardeniers, A. van den Berg, *Electrophoresis*, **24**, 162 (2003).

62. C.Y. Lee, G.B. Lee, L.M. Fu, K.H. Lee, R.J. Yang, *J. Micromech. Microeng.* **14**, 1390 (2004).

63. G.B. Lee, L.M. Fu, C.H. Lin, C.Y. Lee, R.J. Yang, *Electrophoresis*, **25**, 1879 (2004).

64. M. Brivio, R.E. Oosterbroek, W. Verboom, A. van den Berg, D.N. Reinhoudt, *Lab. Chip*, **5**, 1111 (2005).

65. L. Zhang, X. Yin, Z. Fang, *Lab. Chip*, **6**, 258 (2006).

66. H. Chen, Q. Fang, X.F. Yin, Z.L. Fang, *Lab. Chip*, **5**, 719 (2005).

67. L. Zhang, X.F. Yin, *J. Chromatogr. A*, **1137**, 243 (2006).

68. M.M. Crain, R.S. Keynton, K.M. Walsh, T.J. Roussel Jr, R.P. Baldwin, J.F. Naber, D.J. Jackson, *Methods Mol. Biol.*, **339**, 13 (2006).

69. T. Ujiie, T. Kikuchi, T. Ichiki, Y. Horiike, *Jpn. J. Appl. Phys.*, **39**, 3677 (2000).

70. H. Nakanishi, H. Abe, T. Nishimoto, A. Arai, *Bunseki Kagaku*, **47**, 361 (1998).

71. R. Jindal, S.M. Cramer, *J. Chromatogr. A*, **1044**, 277 (2004).

72. F. Kitagawa, T. Tsuneka, Y. Akimoto, K. Sueyoshi, K. Uchiyama, A. Hattori, K. Otsuka, *J. Chromatogr. A*, **1106**, 36 (2006).

73. A.Y Fu, H.P. Chou, C. Spence, F.H. Arnold, S.R. Quake, *Anal. Chem.*, **74**, 2451 (2002).

74. E.T. Lagally, J.R. Scherer, R.G. Blazej, N.M. Toriello, B.A. Diep, M. Ramchandani, G.F. Sensabaugh, L.W. Riley, R.A. Mathies, *Anal. Chem.*, **76**, 3162 (2004).

75. D.S. Zhao, B. Roy, M.T. McCormick, W.G. Kuhr, S.A. Brazill, *Lab. Chip*, **3**, 93 (2003).

76. X. Zhu, G. Liu, Y. Xiong, Y. Guo, Y. Tian, *J. Phys. Conf. Series*, **34**, 875 (2006).

77. F.Q. Nie, M. Macka, L. Barron, D. Connolly, N. Kent, B. Paull, *Analyst*, **132**, 417 (2007).

78. R. Pethig, J.P.H. Burt, A. Parton, N. Rizvi, M.S. Talary, J.A. Tame, *J. Micromech. Microeng.*, **8**, 57 (1998).

79. M.A. Roberts, J.S. Rossier, P. Bercier, H. Girault, *Anal. Chem.*, **69**, 2035 (1997).

80. J.S. Rossier, A. Schwarz, F. Reymond, R. Ferrigno, F. Bianchi, H.H. Girault, *Electrophoresis*, **20**, 727 (1999).

81. I. Nikcevic, S.H. Lee, A. Piruska, C.H. Ahn, T.H. Ridgway, P.A. Limbach, K.R. Wehmeyer, W.R. Heineman, C.J. Seliskar, *J. Chromatogr. A*, **1154**, 444 (2007).

82. M.T. Koesdjojo, Y.H. Tennico, V.T. Remcho, *Anal. Chem.*, **80**, 2311 (2008).

83. M.A. Northrup, M.T. Ching, R.M. White, R.T. Watson. *DNA amplification in a microfabricated reaction chamber*. In *Transducer '93, Seventh International Conference on Solid State Sensors and Actuators*, Yokohama, Japan, 2 (1995).

84. Y.S. Shin, K. Cho, S.H. Lim, S. Chung, S.J. Park, C.L. Chung, D.C. Han, J.K. Chang, *J. Micromech. Microeng.*, **13**, 768 (2003).

85. A. Muck, A. Svatos, *Talanta*, **74**, 333 (2007).

86. J.N. Lee, C. Park, G.M. Whitesides, *Anal. Chem.*, **75**, 6544 (2003).

87. C. Liu, J. Chen, *Sepu*, **23**, 63 (2005).

88. B. Wang, L. Chen, Z. Abdulali-Kanji, J.H. Horton, R.D. Oleschunk, *Langmuir*, **19**, 9792 (2003).

89. S. Hjertén, *J. Chromatogr.*, **347**, 191 (1985).

90. D. Belder, A. Deege, F. Kohler, M. Ludwig, *Electrophoresis.*, **23**, 3567 (2002).

91. Y.W. Lin, M.J. Huang, H.T. Chang, *J. Chromatogr. A*, **1014**, 47 (2003).

92. N.J. Munro, A.F.R. Huhmer, J.P. Landers, *Anal. Chem.*, **73**, 1748 (2001).

93. B.C. Giordano, E.R. Copeland, J.P. Landers, *Electrophoresis*, **22**, 334 (2001).

94. J.C. Sanders, M.C. Breadmore, Y.C. Kwok, K.M. Horsman, J.P. Landers, *Anal. Chem.*, **75**, 986 (2003).

95. S. Lai, X. Cao, L.J. Lee, *Anal. Chem.*, **76**, 1175 (2004).

96. S.L.R. Barker, M.J. Tarlov, H. Canavan, J.J. Hickman, L.E. Locascio, *Anal. Chem.*, **72**, 4899 (2000).

97. S.L.R. Barker, D. Ross, M.J. Tarlov, M. Gaitan, L.E. Locascio, *Anal. Chem.*, **72**, 5925 (2000).

98. Y. Liu, J.C. Fanguy, J.M. Bledsoe, C.S. Henry, *Anal. Chem.*, **72** 5939 (2000).

99. K.W. Ro, W.J. Chang, H. Kim, Y.M. Koo, J.H. Hahn, *Electrophoresis*, **24**, 3253 (2003).

100. D. Wu, Y. Luo, X. Zhou, Z. Dai, B. Lin, *Electrophoresis*, **26**, 211 (2005).

101. M. Pumera, J. Wang, E. Grushka, R. Polsky, *Anal. Chem.*, **73**, 5625 (2001).

102. H. Qu, H. Wang, Y. Huang, W. Zhong, H. Lu, J. Kong, P. Yang, B. Liu. *Anal. Chem.*, **76**, 6426 (2004).

103. X. Jiang, Q. Xu, S.K.W. Dertinger, A.D. Stroock, T.M. Fu, G.M. Whitesides, *Anal. Chem.*, **77**, 2338 (2005).

104. R.C. Gunawan, E.R. Choban, J.E. Conour, J. Silvestre, L.B. Schook, H.R. Gaskins, D.E. Leckband, P.J.A. Kenis, *Langmuir*, **21**, 3061 (2005).

105. A.J. Wang, J.J. Xu, H.Y. Chen, *J. Chromatogr. A.*, **1107**, 257 (2006).

106. X.G. Du, Z.L. Fang, *Electrophoresis*, **26**, 4625 (2005).

107. G.B. Lee, C.H. Lin, K.H. Lee, Y.F. Lin, *Electrophoresis*, **26**, 4616 (2005).

108. Y. Liu, C.S. Henry, *Methods Mol. Biol.*, **339**, 57 (2006).

109. L. Brown, T. Koerner, J.H. Horton, R.D. Oleschuk, *Lab. Chip*, **6**, 66 (2006).

110. S.C. Jacobson, R. Hergenröder, L.B. Koutny, J.M. Ramsey, *Anal. Chem.*, **66**, 2369 (1994).

111. J.P. Kutter, S.C. Jacobson, N. Matsubara, J.M. Ramsey, *Anal. Chem.*, **70** 3291 (1998).

112. S. Constantin, R. Freitag, D. Solignac, A. Sayah, M.A.M. Gijs, *Sensors & Actuators B: Chem.*, **78**, 267 (2001).

113. S.A. Soper, A.C. Henry, B. Vaidya, M. Galloway, M. Wabuyele, R.L. McCarley, *Anal. Chim. Acta*, **470**, 87 (2002).

114. G. Ocvirk, E. Verpoorte, A. Manz, M. Grasserbauer, H.M. Widmer, *Anal. Meth. Instrument*, **2**, 74 (1995).

115. B. He, N. Tait, F. Regnier, *Anal. Chem.*, **70** 3790 (1998).

116. B. He, F. Regnier, *J. Pharm. Biomed. Anal.*, **17**, 925 (1998).

117. C. Ericson, J. Holm, T. Ericson, S. Hjertén, *Anal. Chem.*, **72**, 81 (2000).

118. C. Viklund, E. Ponten, B. Glad, K. Irgum, P. Horsted, F. Svec, *Chem. Mater.*, **9**, 463 (1997).

119. C. Yu, F. Svec, J.M.J. Fréchet, *Electrophoresis*, **21**, 120 (2000).

120. T. Rohr, E.F. Hilder, J.J. Donovan, F. Svec, J.M.J. Fréchet, *Macromolecules*, **36**, 1677 (2003).

121. T. Rohr, D.F. Ogeltree, F. Svec, J.M.J. Fréchet, *Adv. Funct. Mater.*, **13**, 265 (2003).

122. T.B. Stachowiak, T. Rohr, E.F. Hilder, D.S. Peterson, M. Yi, F. Svec, J.M.J. Fréchet, *Electrophoresis*, **24**, 3689 (2003).

123. F. Svec, C. Yu, T. Rohr, J.M.J. Fréchet, in: J.M. Ramsey, A. van den Berg (Eds.), *Micro total analysis systems*, Dordrecht: Kluwer (2001).

124. F. Svec, J.M.J. Fréchet, E.F. Hilder, D.S. Peterson, T. Rohr, in: Y. Baba, A. van den Berg (Eds.), *Micro total analysis systems*, Dordrecht: Kluwer (2002).

125. M.C. Breadmore, S. Shrinivasan, K.A. Wolfe, M.E. Power, J.P. Ferrance, B. Hosticka, P.M. Norris, J.P. Landers, *Electrophoresis*, **23**, 3487 (2002).

126. M.C. Breadmore, S. Shrinivasan, J. Karlinsey, J.P. Ferrance, P.M. Norris, J.P. Landers, *Electrophoresis*, **24**, 1261 (2003).

127. N. Ishizuka, H. Minakuchi, K. Nakanishi, N. Soga, H. Nagayama, K. Hosoya, N. Tanaka, *Anal. Chem.*, **72**, 1275 (2000).

128. B. Vaidya, S.A. Soper, R.L. McCarley, *Analyst*, **127**, 1289 (2002).

129. A.C. Henry, T.J. Tutt, M. Galloway, Y.Y. Davidson, C.S. McWhorter, S.A. Soper, R.L. McCarley, *Anal. Chem.*, **72**, 5331 (2000).

130. F. Xu, M. Jabasini, Y. Baba, *Electrophoresis*, **23**, 3608 (2002).

131. L. Bousse, A. Kopf-Sill, J.W. Parce, *International Conference on Solid State Sensors and Actuators (Transducers)*, June 16–19, Chicago, IL (1997).

132. A. Manz, H. Becker, *International Conference on Solid State Sensors and Actuators (Transducers),* June 16–19, Chicago, IL (1997).

133. B.E. Slentz, N.A. Penner, F. Regnier, *J. Sep. Sci.*, **25**, 1011 (2002).

134. W.N. Vreeland, A.E. Barron, *Curr. Opin. Biotechnol.*, **13**, 87 (2002).

135. K. Sato, M. Yamanaka, H. Takahashi, M. Tokeshi, H. Kimura, T. Kitamori. *Electrophoresis.* **23**, 734 (2002).

136. L. Ceriotti, N. F. de Rooij, E. Verpoorte, *Anal. Chem.*, **74**, 639 (2002).

137. K. Swinney, D.J. Bornhop, *Electrophoresis*, **23**, 613 (2002).

138. L.M. Fu, R.J. Yang, G.B. Lee, *Electrophoresis*, **23**, 602 (2002).

139. X.X. Chen, H.K. Wu, C.D. Mao, G.M. Whitesides, *Anal. Chem.*, **74**, 1772 (2002).

140. T. Thorsen, S.J. Maerkl, S.R. Quake, *Science*, **298**, 580 (2002).

141. I. Gitlin, A.D. Stroock, G.M. Whitesides, A. Ajdari, *Appl. Phys. Lett.*, **83**, 1486 (2003).

142. R.H. Liu, J. Yang, R. Lenigk, J. Bonanno, P. Grodzinski, *Anal. Chem.*, **76**, 1824 (2004).

143. K.A. Shaikh, K.S. Ryu, E.D. Goluch, J.M. Nam, J.W. Liu, S. Thaxton, T.N. Chiesl, A.E. Barron, Y. Lu, C.A. Mirkin, C. Liu, *Proc. Natl. Acad. Sci. U.S.A.*, **102**, 9745 (2005).

144. B.Y. Park, M.J. Madou, *Electrophoresis*, **26**, 3745 (2006).

145. B.M. Taff, J. Voldman, *Anal. Chem.*, **77**, 7976 (2005).

146. K.T. Liao, C.M. Chenb, H.J. Huang, C.H. Lin, *J. Chromatogr. A*, **1165** 213 (2007).

147. R. Zhong, D. Liu, L. Yu, N. Ye, Z. Dai, J. Qin, B. Lin, *Electrophoresis*, **28**, 2920 (2007).

148. Y. Shi, P.C. Simpson, J.R. Scherer, D. Wexler, C. Skibola, M.T. Smith, R.A. Mathies, *Anal. Chem.*, **71**, 5354 (1999).

149. M. Stjernstrom, J. Roeraade, *J. Micromech. Microeng.* **8**, 33 (1998).

150. A. Berthold, L. Nicola, P.M. Sarro, M.J. Vellekoop, *Transducers 99*, Sendai, Japan (1999).

151. Z.L. Huang, J.C. Sanders, C. Dunsmor, H. Ahmadzadeh, J.P. Landers, *Electrophoresis*, **22**, 3924 (2001).

152. T. Ito, K. Sobue, S. Ohya, *Sensors & Actuators B*, **81**, 187 (2002).

153. Z.J. Jia, Q. Fang, Z.L. Fang, *Anal. Chem.*, **76**, 5597 (2004).

154. S. Satyanarayana, R.N. Karnik, A.J. Majumdar, *Microelectromech Sys.*, **14**, 392 (2005).

155. M.M.R. Howlader, S. Suehara, T. Suga, *Sensors & Actuators A*, **127**, 31(2006).

156. A. Satoh, *Sensors & Actuators A*, **72**, 160 (1999).

157. A. Weinert, P. Amirfeiz, S. Bengtsson, *Sensors & Actuators, A*, **92**, 214 (2001).

158. M.J. Wild, A. Gillner, R. Poprawe, *Sensors & Actuators A*, **93**, 63 (2001).

159. H.K. Wu, B. Huang, R.N. Zare, *Lab. Chip*, **5**, 1393 (2005).

160. A.Y.N. Hui, G. Wang, B.C. Lin, W. T. Chan, *Lab. Chip*, **5**, 1173 (2005).

161. A.A. Yussuf, I. Sbarski, J.P. Hayes, M. Solomon, N. Tran, *J. Micromech. Microeng.*, **15**, 1692 (2005).

162. M.T. Blom, E. Chmela, J.G.E. Gardeniers, J.W. Berenschot, M. Elwenspoek, R. Tijssen, A. van den Berg, *J. Micromech. Microeng.*, **11**, 382 (2001).

163. R.T. Kelly, T. Pan, A.T. Woolley, *Anal. Chem.*, **77**, 3536 (2005).

164. A.V. Pattekar, M.V. Kothare, *J. Micromech. Microeng.*, **13**, 337 (2003).

165. E.S. Lee, D. Howard, E. Liang, S.D. Collins, R.L. Smith. *J. Micromech. Microeng.*, **14**, 535 (2004).

CHAPTER 3

INSTRUMENTATION OF NANOCHROMATOGRAPHY AND NANOCAPILLARY ELECTROPHORESIS

3.1. INTRODUCTION

Basically, the efficiency, selectivity, robustness, speed, detection, and reproducibility of all the analytical techniques are the inherent characteristics of their paraphernalia. All parts of the instruments are important as they play crucial roles to achieve the nanolevel analyses. Nanoanalysis by chromatography and capillary electrophoresis is an emerging field in separation science and, hence, an introduction to the paraphernalia is very important before proceeding with the experimental work. In nanochromatography the most important aspects to be considered are mobile phase reservoirs, tubings, pumps, injectors, columns, detectors, and recorders. In the case of nanocapillary electrophoresis, solvent reservoirs, injection ports, detectors, and recorders are responsible for providing detection at trace levels. Nano detection in samples of low volume or having low concentrations of ingredients is difficult. Therefore, knowledge of the configuration of the sample preparation device is also very important to achieve the desired task. It is important to mention that nanoanalyses have been achieved by using normal liquid chromatographic, gas chromatographic, and capillary electrophoretic instruments involving capillary columns, but these are not discussed here as information on their instrumentation is available in many articles and books [1–7] and interested

Nanochromatography and Nanocapillary Electrophoresis. By Ali, Aboul-Enein, and Gupta
Copyright © 2009 John Wiley & Sons, Inc.

readers can consult them. This chapter is dedicated to the discussion on microchip-based paraphernalia required for trace analyses in the nano world of separation science. Nanochromatographic and nanocapillary electrophoretic instruments are not yet fully developed; however, attempts have been made to explain the instrumentation available today.

3.2. NANOLIQUID CHROMATOGRAPHY (NLC)

Great efforts have been made to develop nanoliquid chromatographic instruments by using theoretical, technical, and methodological studies. Nanolevel separations have been achieved through nano-HPLC but few papers are available on this subject using other modalities of liquid chromatography, such as capillary electrochromatography (CEC) and micellar electrokinetic chromatography (MEKC). Generally, nano separations by these modalities are achieved using tubular or packed capillaries of 10 to 100 μm diameter with an MS detector. Few papers were found on microchip-based nano-HPLC, which is truly a nano-HPLC technique and is the future of nanolevel separations. Therefore, this chapter describes instrumentation for chip-based nano-HPLC. An outlook and schematic representation of nano-HPLC instruments are shown in Figs. 3.1 and 3.2, respectively.

Basically, nano-LC and nano-CE are not yet fully developed but are still growing. Therefore, few papers are available in the literature on nanoanalyses using real nano-HPLC. However, some workers have attempted to develop nano-HPLC machines. Szekely and Freitag [8] designed and described a chip-based analytical system and the monolithic stationary phase was developed in the separation channel of the chip for pressure- or voltage-driven nanochromatography. The multifunctional connection and support unit having optical probes, reservoirs, and high pressure and voltage connectors were

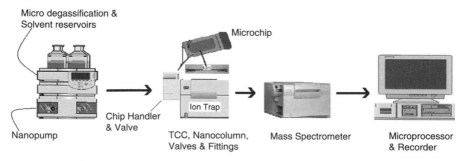

Figure 3.1 A complete chip-based nano-HPLC machine.

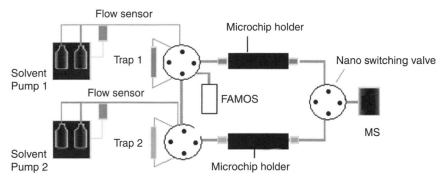

Figure 3.2   A schematic representation of nano-HPLC.

also developed. The authors reported the chips were capable of bearing pressure of 130 to 150 bar. The authors tested this nano-HPLC system in terms of its fluid dynamics under pressure- and voltage-driven conditions. A schematic presentation of the chip holder, reservoirs, flow control, high pressure connections, and optical detection is shown in Fig. 3.3 [8]. In this system, the chip holder/interface was made from Teflon. Gold electrodes

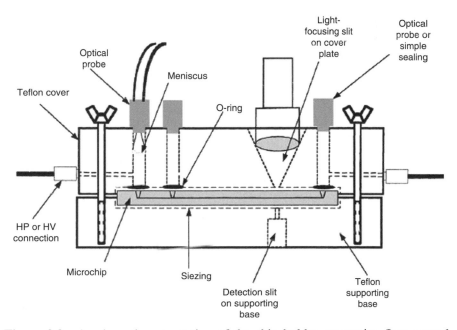

Figure 3.3   A schematic presentation of the chip holder, reservoirs, flow control, high pressure connections, and optical detection in NLC [8].

were inserted from the side of the reservoirs and noninvasive optical flow control unit and connectors were integrated. A second holder was used for washing, which is attached to the HPLC pump; with all connections to the chip holder including high voltage power supply.

Nanostream Inc. USA and Agilent, USA are the main manufactures of LC-chips. Nanostream Inc. has introduced chips made of brio cartridges of polymeric material with 24 bead-packed parallel microfluidic columns by which 24 samples can be analyzed simultaneously. On the other hand Agilent introduced a chip-based fritless LC column with electrospray ionization (ESI) nozzle. The chips are made of two polyimide layers bonded together. The columns are packed with C_{18} material by using a tapered outlet. The samples can be injected from off-chip by using an injection valve attached to the back of the chip. Various components of a nano-HPLC are discussed in the following sections.

3.2.1. Mobile Phase Reservoirs

Basically, no special devices have been developed to handle mobile phases in nano-HPLC. However, our experience dictates that reservoirs small in volume and of high quality glass are preferred for this purpose. Solvent containers should be air tight and free from any contamination. The use of helium gas, through a sparging device, may be beneficial for degasification of solvents in nano-HPLC, which can improve check valve reliabilities, especially at nano flow, and diminish baseline noise in UV detection. Besides, each reservoir should be equipped with a shutoff valve for efficient helium consumption [9].

3.2.2. Mobile Phases and Flow Calibration

The mobile phase is the heart of separation in liquid chromatography and, hence, solvents must be of HPLC grade and should be degassed before their use. Commercial available nano-HPLC systems have online degassers in their pumping units. The preferred organic solvent is acetonitrile instead of classical methanol due to the viscous nature of the latter, which creates problems at high pressure with nanolevel flow rates. Methanol viscosity is 0.60 cP at 20°C, while that of acetonitrile is 0.37 cP. Besides, high grade purity water is also a useful mobile phase in nano-HPLC, which can be used in various combinations of other solvents. Ethanol and propanol have also been used in nano-HPLC. Reproducibility and flow accuracy of the mobile phases in a nano-HPLC machine is controlled by the new microchip flow sensor. This sensor controls precise flow rates, which is characterized by high precision,

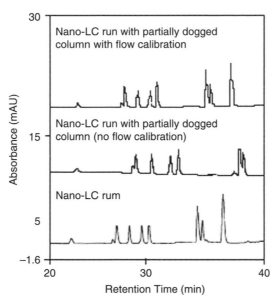

Figure 3.4   The restored chromatography using flow calibration [10].

digital intelligence, and excellent reliability. The noninvasive and measuring of nano flow rates are independent of mobile phase composition. The mobile phase flow calibration provides constant flow delivery with zero chromatographic dispersion. Figure 3.4 indicates microflow sensing and calibration compensation for anomalies in column and system performance to ensure consistent chromatographic reproducibilities [10].

3.2.3. Mobile Phase Tubings

In nano-HPLC, high grade tubings are required and must be made of steel or some other inert material like polyether ether ketone (PEEK). To the best of our experience PEEK tubings are useful to achieve excellent performance in nano-HPLC. The diameter of the tubing should be as small as possible, preferably in the micrometer range. The internal surface should be smooth for providing constant laminar flow of mobile phase. The lowest possible tolerance, on concentricity of tubing and good alignment of the bores is effective to achieve the nanoscale separation. Reduction in void volume may be achieved by minimizing the number of connections and use of low volume accessories such as online filters, spillers, and unions. To achieve constant laminar flow, the connections of tubings must be with minimum connection gaps.

3.2.4. Solvent Delivery Pump

Nano-HPLC needs reliable pumps especially for providing constant flow rate with no variation. The pumps are capable of delivering a flow rate ranging from 25 to 4,000 nL/min nano-HPLC pumps have gradient flow delivery systems capable of high reproducibility at low flow rates. Such a low flow rate is generated by applying nano flow splitting technique. For this purpose, a newly designed, noninvasive microchip-based nanoflow sensor (Fig. 3.5) has been used in this technique. These types of flow devices were developed by Dionex Corporation, USA. Other commercially available pumps for nano-HPLC have appeared on the market, which use compressed air as the pressure source and a flow sensor to provide feedback control for generating exact and precise nL/min flow rates [11]. Scivex, USA introduced a confluent NFM (nanofluidic module), which delivers isocratic solvents at extremely low flow rates, which is compatible with sample loading into an MS, lab-on-a-chip and nanoliquid chromatography. It is important to note that a quick-split automated flow splitter can convert a standard HPLC system into nano-HPLC capable of delivering 100 nL/min flow rate. The flow splitting pumps generate highly reproducible flow rates in nano-HPLC with fast gradient equilibration times.

Xie et al. [9] reported on-chip generation of gradient elution by using electrochemical pumping in LC-ESI, which contains two 3 μL on-chip solvent reservoirs, two electrochemical pumps, and an ESI nozzle. Solvent pumping was controlled galvanostatically at a flow rate of 200 nL/min. The authors reported a higher electrical current was used to increase flow rate or to achieve

Figure 3.5 Noninvasive microchip-based nanoflow sensor.

pumping at higher back pressures. For example, pumping at 200 psi back pressure was demonstrated using a current of 800 pA. The authors also claimed the generation of electrically controlled on-chip gradient without any external fluidic connections to the chip. They also observed continuous generation of gradients for 30 min at a total flow rate of 100 nL/min. Furthermore, the authors advocated this pumping device as capable of meeting the general requirements of LC applications. Recently, Nie et al. [12] introduced a robust, compact, on-chip, electroosmotic micropump (EOP) for microflow analysis, which may be useful in nano-HPLC. The authors fabricated a chip (15 × 40 × 2 mm) made of poly(methyl) methacrylate (PMMA). The schematic representation of the fabrication process for an electroosmotic pump is shown in Fig. 3.6.

Electroosmotic pumps lack mechanical parts and specific localization in the manifold, producing an even electroosmotic flow. Besides, the flow in interconnected and branched channels can be controlled by switching voltages only. Just two decades ago electroosmotic pumps were attractive and feasible ways for mobile phase flow into microfluidic devices [13] but in the 1990s the conventional pumps available showed a major problem with the high pressures

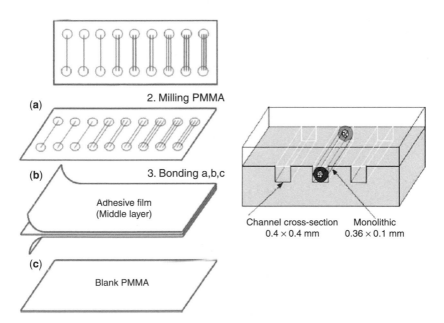

Figure 3.6 A schematic representation of the fabrication process for an electroosmotic pump. (Left) (a) milled PMMA chip, (b) channels containing fixed capillaries, and (c) blank PMMA top plate attached. (Right) A schematic view of the monolithic column inside the channel sections [12].

necessary for transport in channels in the order of 1 to 20 μm i.d. and 1 to 10 m length. Therefore, attempts were made to carry out development of micropump systems [14,15] for microflow systems. Manz et al. [16,17] designed and developed valves, pumps, and detection systems, in miniaturized flow injection analysis, stacked modular devices in silicon and plexiglass for suppressing limitations of the integration.

3.2.5. Sample Injector

The sample injection is the most important and critical issue in nano-HPLC, which requires quality, reproducibility, and flexibility of sample injection valves. A low dispersion injection method is used in nano-LC to achieve sharp peaks with high quality chromatographic resolution. Micro-autosamplers, with loading capacity of 1 μL, are the ideal ones for nano-HPLC. Normally, sample injectors have two positions, that is, load and inject, which allow sample loading and injection, respectively. The injectors used in the normal HPLC pumps may be used to load a low amount of sample (microliter). Rheodyne, USA launched a new microscale injector (Model 7520), which may be useful for nano-HPLC by injecting a 0.2 μL sample via coupling to accessory chambers. The internal configuration of the injector is shown in Fig. 3.7, which shows its configuration with flow direction of the mobile phase. The same company designed an automatic nanoscale injector (Model MX), which is ideal for nano-HPLC. The configuration of this injector includes a 10 μL internal injection loop, low dispersion 0.10 mm flow

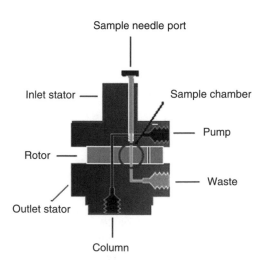

Figure 3.7   Internal configuration of an injector (Model 7520) [18].

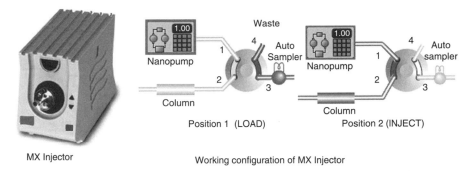

Position 1 (LOAD)   Position 2 (INJECT)

MX Injector   Working configuration of MX Injector

Figure 3.8   The MX injector and its working configuration [18].

passages, and operation by either push button or complete remote control. The special feature of this injector comprises the loading of sample during equilibration and injection unattended; when the system is ready just by connecting the valve to a single contact closure (two wires). The MX injector and its working configuration are shown in Fig. 3.8 [18]. The configuration and working capacity of the MX injector are given in Table 3.1, which indicates various features of the injector, such as maximum pressure, flow passages, operating temperature, power requirements, and remote control communication. Dionex, USA also developed a sample injection device using a specially designed micro-autosampler (up to 1 μL). Ye et al. [19] developed tubing set up for an autosampler valve to achieve nanodetection of phenols. Groton, USA launched a new version of automated reactor sampling (ARS) for sampling in nanoliquid chromatography with the facility of multiplexation with others to improve throughput. Shen et al. [20] developed a microfluidic chip-based sequential injection system for the detection of butyl rhodamine B. The sample injection system was slotted with a vial array for holding sample and reagents using six micropipette tips with slotted ends.

TABLE 3.1   Configuration and Working Capacity of the MX Injector

Parameters	Specifications
Maximum pressure	5000 psi (345 bar)
Flow passages	0.1 mm (0.004 in.) diameter
Power requirements	100–240 VAC, 50/60 Hz
Remote control communication	One-line contact closure
Operating temperature	4–40°C
Dimensions	10.2 cm H × 7.6 cm W × 12.7 cm D

Chmela et al. [21] described a pressure-driven injection system for an ultra-flat chromatographic microchannel, compatible with shallow microchannels with a very large aspect ratio (1 μm deep and 1000 μm wide). The authors claimed this sort of system provides high loadability and low sample dispersion, and, hence offer potential advantages in the miniaturization of liquid chromatography. Furthermore, the authors carried out computational fluid dynamics simulations to predict the flow profiles and the transport of solute in the system and justify the injection principle. Based on the simulation results, a prototype integrated into a chip for hydrodynamic chromatography has been realized and tested experimentally by these workers. The performance of the device was advocated as satisfactory and the results were in qualitative agreement with the numerical models.

3.2.6. Separation Chips

Separation capacity, reproducibility, ruggedness, and efficiency of a nano-HPLC depend on dimension and packing material microchips. Therefore, the chip is the most important and crucial part of a nano-HPLC system. The selection of chip dimension and packing material depends on the type of applications and experimental conditions. An experienced person can select suitable chips more efficiently and precisely. But chip technology is not fully developed and there are some limitations in the selection of chip dimensions and packing materials. However, the best chip selection can dramatically cut time, save costs, and increase productivity while improving analytical results. The columns engraved on microchips (75 μm, i.d.) are packed or coated with C_8 or C_{18} silica gels, the packing materials normally used in conventional HPLC. Normally, the particle sizes of silica gels are 2.1, 3.5, or 5.1 μm. The most important factors reducing column performance are void volumes, nonspecific interactions, and disruption of the laminar flow. Therefore, to achieve nanoscale separation high quality packing materials, integrated connection capillaries, PEEKsil (fused silica lined PEEK for column body and tubings) are required. Besides, all surfaces of the packing material is deactivated, that is, the silonol group on the fused silica gel is chemically deactivated. The stainless steel fibers of the frit are gold plated and covalently deactivated. In nano-HPLC, chips are fully optimized by minimizing system dead volume.

The column length and inner diameter are the two most important features required in column generation on microchips. The column separation capacity is measured in terms of number of plates, which is proportional to column length. But back pressure and analysis time are raised proportionally as column length increases. For gradient separations, column length is less a factor for resolution as separation is controlled by gradient rather than length of the column. However, the optimum length of nano-HPLC columns ranged from 1 to 5 cm. Narrow bore columns are needed in trace analysis, which require less sample and mobile phase. It should be noted that on-chip columns usually have rectangular cross-sections, unlike the conventional cylindrical columns. Due to these facts, nano-HPLC is sometimes also called nano-bore HPLC and nano-scale HPLC [22]. Normally, nano-HPLC columns on chips have 20 to 100 μm as the internal diameter. The internal diameters of columns used in different modalities of liquid chromatography are given in Table 3.2. The column switching techniques are also useful and effective in nano-HPLC. Newly developed monolithic columns, with column switching module, have a wide range of applications with good efficiencies and reproducibilities, which can be used in nano-HPLC. The nano-scale MX six-port switching valve (Stainless Steel MX7980-000) or an analytical-scale MX six-port switching valve (Stainless Steel MX7900-000 or Biocompatible MX9900-000) are available devices for automated column back flushing [10].

In spite of the importance of nano-HPLC, on-chip pressure-driven nano-HPLC technology is not fully developed but progress is being made [23–33]. One factor is the lack of available technologies to integrate various components of a nano-HPLC system, especially the separation column [24,25]. The packing of an on-chip column with beads seems to be desirable and straightforward with few exceptions [26]. These include open tubular [27], surface-activated micromachined posts [28], continuous monolithic bed formed by in situ polymerization [29], and column coated with nanoparticles [30]. Agilent, USA has provided a chip-based fritless HPLC column capable of coupling with an electrospray ionization (ESI) nozzle [23,31]. The chip is fabricated of polyimide layers while the ESI nozzle is fabricated by laser ablative trimming of the bonded chip. Xie et al. [32] developed the most complete

TABLE 3.2   The Different Types of Columns and Required Mobile Phase Flow Rates Used in Different Modalities of HPLC[a]

Columns	Internal Diameters	Flow Rates
Open tubular LC	<25 μm	<25 nL/min
Nano-HPLC	25–100 μm	25–4,000 nL/min
Capillary-HPLC	100–1,000 μm	0.4–200 μL/min
Micro-HPLC	1.0–2.1 mm	50–1,000 μL/min
Normal HPLC	4.0–5.0 mm	1.0–10.0 mL/min
Preparative HPLC	>10 mm	>20 mL/min

[a]Ref. 16.

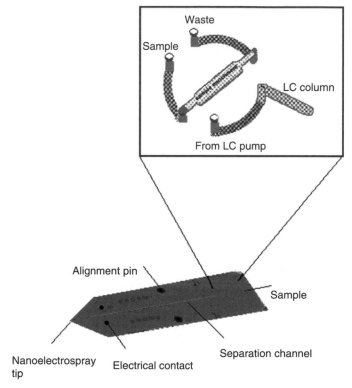

Figure 3.9   A schematic representation of an LC-microchip.

nano-HPLC chip that has on-chip gradient pumping, sample injection, column, and electrospray nozzle, with a column length of 1.0 cm. ACE (www.ace-hplc.com) introduced nano-HPLC chip-based columns of 75 to 100 μm internal diameter with 100 Å pore size of the packing material. To familiarize readers with the chip-based separation device, a schematic representation of a microfluidic chip integrated with a 30 nL volume precolumn, a 4.5 cm length analytical column (75 × 50 μm cross-section channel) and a 10 μm i.d. nano electrospray emitter is shown in Fig. 3.9.

3.2.7. Detectors

Nanoanalyses relay on the most sensitive, efficient, and reproducible detectors. The art of hyphenation of the detectors with nano-HPLC is the most important and crucial aspect in separation science at the trace level. In spite of the use of many detectors in nano-LC, the sensitivity of detection at the nanolevel is still a challenging job. However, this task may be achieved by using very

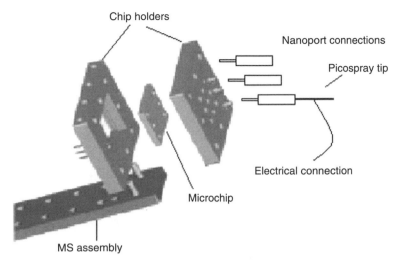

Figure 3.10   A schematic representation of chip-based MS [46].

sensitive detectors, such as UV-visible, conductivity, fluorescent, refractive index, atomic absorption spectroscopy (AAS), atomic emission spectroscopy (AES), atomic fluorescence spectrometry (AFS), inductively coupled plasma (ICP), mass spectrometry (MS), nuclear magnetic resonance (NMR), inductively coupled plasma-mass spectrometry (ICP-MS) and time-of-flight mass spectrometry (TOF-MS) [35–41]. The hyphenation of these detectors with nano-HPLC improves sensitivity greatly. Many hyphenation methods have been reported in the literature for normal LC, which may be also used for nano-world machines. Details are provided in Chapter 4 of this book. Basically, nano-HPLC deals with small amounts (20–50 nL) of sample and, hence, the sensitivity is not very high. Therefore, some devices have been coupled with nano-HPLC, which increase sensitivity in detection. In the case of UV-visible detection, an increase in volume of the detector cell is useful to raise the detection level. Besides, column focusing [42,43] and 2-D separation [44] methods have been used to avoid this drawback. Nowadays, MS detectors are available based on chip technology, which can be coupled with chip-based nano-HPLC [45,46] A schematic representation of a microchip-based mass spectrometer detector is shown in Fig. 3.10 indicating the alignment of the chip in the MS unit.

3.2.8. Recorder

The recorder also plays an important role in analysis by printing the chromatograms. Nowadays, all nano-HPLC instruments are computer based and data are processed through different software. The most effective nano-HPLC software are Millennium 2000, Sparklink, Agilent ChemStation A 10.02, Chromleoon, and UltiChrom. These are capable of sensing even a single signal and can record it in the form of a chromatogram. The facilities of chromatogram expansion are also available by which a small peak can be expanded, integrated, and observed easily. Moreover, these software include many other facilities like comparison of standard and sample chromatograms, nanoscale quantitation, calculations of methods of error, etc., to process the analytical data. Shih et al. [47] described on-chip temperature gradient interaction chromatography. The system consisted of a parylene high pressure liquid column, an electrochemical sensor, a heater, and a thermal isolation structure for on-chip temperature gradient interaction chromatography applications. The column was 8 mm long, 100 μm wide, 25 μm high, packed with 5 μm sized C_{18} silica gel.

3.2.9. Sample Preparation Units

Trace analyses in unknown matrices are challenging owing to the presence of many impurities, which interfere with the separation and detection. Normally, nanoseparation is required for samples of low volume or poor ingredient concentrations and under such conditions sample preparation becomes crucial especially in biological and environmental matrices. In addition, impurities in biological and environmental matrices make nanoanalysis more complex. In view of these facts, nanoanalysis requires optimized sample preparation. Many online methods are available for sample preparation in normal HPLC, which may also be used in nano-HPLC. Details of sample preparation methodologies are discussed in Chapter 3. However, the hyphenation of sample preparation units with chromatography is discussed briefly here. Veuthey et al. [48] described an online solid phase extraction unit to achieve nanoanalysis of drugs in biological samples. Injection of the sample is carried out directly onto the extraction support with and after the extraction, the valve is switched on, and analytes are transferred to the column with the eluting mobile phase followed by extraction support at re-equilibration. The method reported is simple and several applications have been published for direct analysis of biofluids. Recently, Pragst [49] reviewed sample preparation by solid phase micro-extraction devices and information furnished in this article may be useful for sample preparation in nanoliquid chromatography. Shen et al. [20] described a microfluidic chip liquid–liquid extraction preconcentration of butyl rhodamine B. The schematic representation of the chip-based microfluidic system is shown in Fig. 3.11. The unit comprises a microfabricated glass lab chip with a 35 mm extraction channel and 134

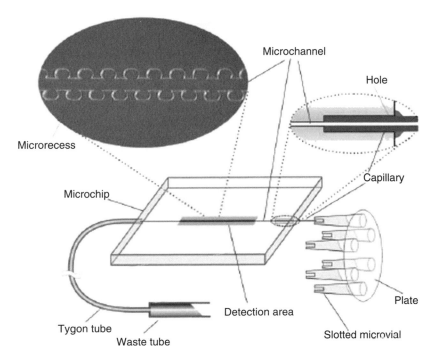

Figure 3.11   A schematic representation of a chip-based microfluidic system [20].

shrunken opening rectangular recesses (100 mm length, 50 mm width, 25 mm deep) arrayed within a 1 mm length on both sides of the middle section of the channel. Many other sample preparation devices are also available based on chip technology [50–53]. Vanhoutte et al. [54] compared the amount injected and UV and MS responses for a phosphate alkylated dGMP/BPADGE adduct. The results are given in Table 3.3, which indicates that nanoliquid chromatography (NLC) is the best modality due to the low amount of sample injection and high UV and MS responses. Besides, the response of MS

TABLE 3.3   A Comparison of Nanocapillary and Conventional LC-MS for the Amount Injected and UV and MS Responses[a]

LC/MS	Volume Injected, nL	Amount Injected, ng	UV Response	MS Response
Nano	3.0	0.135	1.0	1.0
Capillary	200.0	9.0	11.0	4.1
Conventional	20000.0	900.0	44.0	2.3

[a]Ref. 56.

TABLE 3.4   The Effect of Chip Implementation on Band Broadening[a]

Flow Rate, μmL/min	Total Microfluidic System	Contribution of the Chip	Total Microfluidic System	Contribution of the Chip
1	39	33	100	49
2	26	23	54	29
3	21	18	39	23
4	18	15	31	20

[a]Ref. 57.

detector is good in comparison to UV. de Boer et al. [55] studied the effect of chip implementation on the band broadening and the results are given in Table 3.4, which indicates that lower flow rates and injection volumes resulted in broader peaks. Furthermore, it may be seen that at an injection volume of 0.1 mL, 85% of the band broadening can be contributed to the microfluidic chip, independent of the flow rate. It is due to the fact that the connections and channels of the chip increased the extra column volume and thus the sample dilution. At larger injection volumes (1 mL), the percentage of band broadening was 60% due to the diffusion at the borders of the sample plug.

3.3. NANOCAPILLARY ELECTROPHORESIS (NCE)

Capillary electrophoresis (CE) is a separation technique involving nanoliter samples in a fused silica capillary of 30 to 100 cm length and 50 to 100 μm diameter. Mostly the detection limit of this configuration is at the microlevel, with the exception of nanodetection and, hence, it is a popular technique in the microworld. Due to high demand for nano-level analyses of samples with low volume and poor concentration, attempts have been made to achieve this task by incorporating a chip in CE, which is called microchip capillary electrophoresis (MCE). In this book we term it nanocapillary electrophoresis (NCE) as it involves sample injection and flow of background electrolyte (BGE) at the nanolevel. The advances in genome, proteomic, and drug development compel scientists towards the development of chip-based CE. Chip-based CE has many advantages, such as high speed, reduced sample volume, and higher separation efficiency [56–61]. The integration of all components of CE on to a chip was developed and called the Lab-on-Chip system, comprising a microchannel network for sample pre- and post-handling, chemical reactions, separation, and detection.

The development of chip-based CE is an innovation in the fields of chemical, clinical, biological, and other analytical sciences. Manz et al. [56] used planar chips technology for miniaturization and integration of separation in capillary electrophoresis on a chip. Normally, the NCE cross-channel chip is made of silica or glass, or other materials such as poly(dimethylsiloxane) (PDMS) with a channel cross-section of 50×20 μm and separation channel length of 5 cm. Among the materials listed above glass is the best one due to its favorable electrical insulation and thermal conductivity, which are required for electrokinetic pumping using electroosmotic flow instead of mechanical pumping. Quartz is the best glass used for this purpose owing to its good transparency in the UV region, which is required in ultraviolet detection. Agilent/Caliper's 2001 Bioanalyzer and Caliper's Automated Fluidics System 90 are chip-based CE instruments. A schematic representation of a chip-based NCE with negative pressure, large volume sample injection is shown in Fig. 3.12 [62]. The instrument contains a syringe pump (SP),

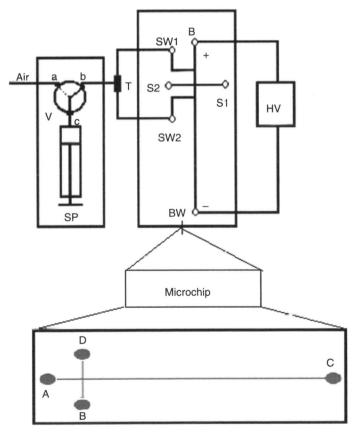

Figure 3.12   A schematic representation of NCE [62].

three-way valve (V), high-voltage power (HV) supply, T-shaped connector (T), buffer solution reservoirs (B and BW), sample containers (S1 and S2), and sample waste reservoirs (SW1 and SW2). In this set up, the chip is operated in sample loading mode and the three-way valve is switched to connect c to b and the pump is used to aspirate 200 μL air from the two sample waste reservoirs. The negative pressure developed is responsible for sample loading. Simultaneously, buffer solutions are also drawn from reservoir B to SW1, and BW to SW2, introducing a large volume sample plug. The experiment is carried out by applying high voltage as in the case of a normal CE machine.

3.3.1. Separation Chip

A microfabricated CE chip has been developed and many workers have reported a promising future for it [56,63]. The microchip is prepared by photolithography and wet chemical procedures using a hydrofluoric acid and nitric acid mixture. Contrarily, dry etching of silica gel has been investigated intensively for oxide contact hole etching. In wet and dry etching the etched lengths are 1 to 2 μm and 5 to 10 μm, respectively [64,65]. These chips may be used for sample preparation, separation, and detection (biochemical sensors, microoptics, and microelectronic devices) for very small samples [66]. Microchannels are obtained on quartz glass using photolithography and wet chemical etching by hydrofluoric acid or a mixture of hydrofluoric acid and nitric acid.

Ujji et al. [67] reported high rate quartz etching with high selectivity using a plasma process, and developed chips of high-aspect-ratio capillaries or submicron structures, which are not possible by conventional wet technology. These authors compared the dry etching technology with the wet process and reported that the dry method was the best approach. They also described the process of quartz microcapillary electrophoresis chip preparation as shown in Fig. 3.13 [67]. The authors used chromium film sputtered on an optically flat quartz plate of 0.5 mm thickness and 20 × 20 mm size, patterned using photolithography and wet etching. The capillary trenches were in fluorine-based inductively coupled plasma. The chromium residue and other chemicals were removed from the chips using 1% hydrogen fluoride and bonded together through four holes. Finally, two chips are pressed together at 1.3 MPa for 24 h to obtain an enclosed channel used as a capillary.

Figure 3.14 [67] shows a design of a micocapillary electrophoresis chip, which is a closely folded separator 7.6 cm long from the sample reservoir to the waste port, engraved onto a 20 × 20 mm^2 quartz plate. A double-T sample injector is upstream of the separator. The etched capillary is rectangular

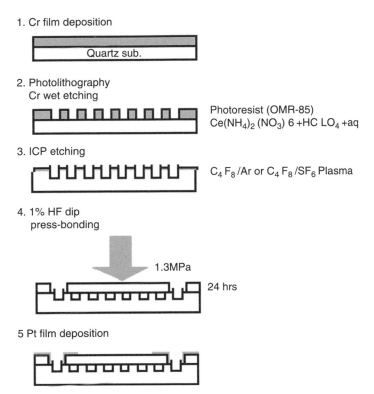

1. Cr film deposition

Quartz sub.

2. Photolithography
   Cr wet etching

Photoresist (OMR-85)
Ce(NH$_4$)$_2$ (NO$_3$) 6 +HC LO$_4$ +aq

3. ICP etching

C$_4$ F$_8$ /Ar or C$_4$ F$_8$ /SF$_6$ Plasma

4. 1% HF dip
   press-bonding

1.3MPa

24 hrs

5 Pt film deposition

Figure 3.13   A schematic representation of the process sequence of a quartz NCE chip [67].

in shape with cross-sections of 30 × 30, 15 × 30, and 30 × 15 μm to investigate the cross-sectional effect on chip performance. The authors reported improved heat dissipation in comparison to normal silica capillaries. The separation of rhodamine B and sulforhodamine dyes was obtained within 20 seconds just applying 2 kV as applied potential with 20,000 as theoretical plates. Furthermore, these authors studied the working characteristics of the chips and reported that glass chips are better than silica ones due to improved heat dissipation. Figure 3.15 indicates plots of current against voltage for 7.6 cm chips of 30 × 30, 30 × 15, and 15 × 30 μm, which indicates almost similar behavior except for the 30 × 30 μm chip, which shows 1.5 times higher electrical current. The authors reported that the current area was independent of cross-sectional area but proportional to the periphery length. Furthermore, the authors reported that the same behavior of 15 × 30 and 30 × 15 μm chips indicates microstripes on their etched sidewall have negligible influence on separation characteristics.

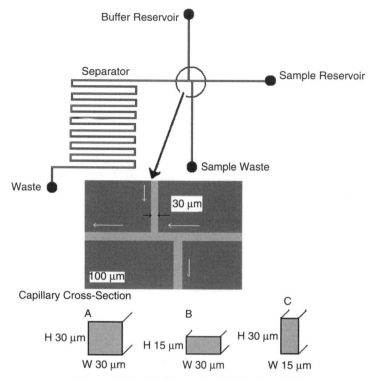

Figure 3.14 Design of an NCE chip [67].

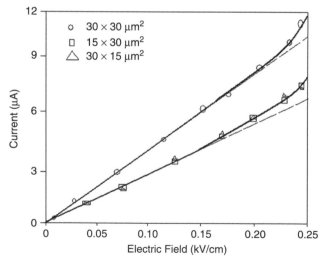

Figure 3.15 I-V characteristics of NCE chip with capillaries of different aspect ratios [67].

3.3.2. Background Electrolyte and Its Reservoirs

There are no special devices used for the background electrolyte (BGE) in NCE and a small vessel is etched on the chip, which is filled with BGE. Schematic representation of the buffer reservoirs (B) can be seen in Fig. 3.12, which indicates the location of the BGE reservoir. Background electrolytes used in conventional CE can be used in NCE effectively. There is no special BGE required in NCE. However, the solvents and reagents should be of HPLC and analytical grades. Normally, phosphate buffers of different pHs and ionic strength are used in NCE. Different organic modifiers such as acetonitrile and methanol may be used to improve resolution. The exhaustive degasification of the BGE is required for reproducible results. Based on our experience, fresh BGEs are most effective for efficiency and reproducibility. The reproducibility of NCE is better than commercial CE as the former modality needs small amounts of BGE, keeping the ionic concentration almost constant. Jackson et al. [68] used phosphate buffer (20 mM, pH 4) and filled in NCE reservoirs. There is no need to discuss the composition of BGE in NCE here since these are given in many review articles [59–61,68–84]. Fu et al. [85] described an experimental and numerical investigation into electrokinetic focusing flow injection for bioanalytical applications with a novel device that integrated two important microfluidic phenomena, that is, electrokinetic focusing and valveless flow switching within multiported microchannels. According to this study a voltage control model achieving electrokinetic focusing, in a pre-focusing sample injection system, allowed the volume of the sample to be controlled. Chang and Yang [86] reviewed recent developments in electrokinetic mixing in microfluidic systems.

3.3.3. Sample Injection Port

Sample injection in NCE is very important for reproducible results with low limits of detection. In spite of some development in NCE very little effort has been made to develop sample injection devices in this technique. Of course sample injection in NCE is a challenging job due to small volume requirement [87]. The controlled injection of small amounts of sample is a prerequisite for successful analysis in NCE. Electrokinetic injection (based on electroosmotic flow) is the preferred method and Jacobson et al. [88] optimized sample injection using this approach. Pinched injection allowing injection in minute quantities [89,90] and double-T shaped fluidic channels [91] have also been used for this purpose. Furthermore, Jacobson et al. [92] used a single high voltage source to simplify instrumentation. Similarly Zhang and Manz [93] developed a narrow sample channel injector to improve resolution and sensitivity. Thomas et al. [94] modified the pinched method for rapid sample loading.

The electrokinetic technique depends on surface properties of channel walls resulting in the bias effect towards different species [95,96], especially when the surface properties are modified by injected samples. Long time injection strategies ranging from 10 to 150 seconds are useful to avoid this problem. Fang et al. [97] reported on the sequential injection sample introduction method in microfluidic NCE, the authors observed reproducible injections using this method. Bai et al. [98] proposed a pressure pinched injection of nanoliter volumes by a multiport injection valve with three syringe pumps. Solignac and Gijs [99] developed hydrodynamical injection by applying a pressure pulse to a membrane on the reservoir. Kaniansky et al. [100] used fixed volume sample loops and four peristaltic micropumps for injection. These authors also described column switching in NCE. Baldock et al. [101] developed a sample injector capable of delivering variable sample amounts to the microdevice with both hydrostatic pumping and syringe pumps. Gai et al. [102] described injection in NCE by hydrostatic pressure in conjunction with electrokinetic force on a microfluidic chip. The hydrostatic pressure was generated by adjusting the liquid level in different reservoirs without any additional pump. Two dispensing strategies using floating and gated injections, coupled with hydrostatic pressure loading, were tested. Futterer et al. [87] described an injection system with dynamic control of the reservoir pressures at the end of each channel. Wu et al. [103] reported a push/pull sample injection method by using push/pull pressure flow via a dual-syringe pump. Cho et al. [104] described a bias-free pneumatic sample injection in NCE, which enabled solution introduction without sampling bias. An aliquate of 10 nL was injected by pneumatic pressure into a hydrophilic separation channel. Buttgenbach and Wilke [105] described a hydrodynamic sample injection procedure in poly(dimethylsiloxane) NCE. The authors advocated a high degree of functional integration in miniaturization. Recently, Zhang et al. [106] proposed an efficient and reliable approach for injecting well-defined, nonbiased picoliter sample plugs into the separation channel of an NCE. The injection was carried out by pressure-driven flow induced via negative pressure, generated by a syringe pump with EOF, and static pressure-driven flows created by differential liquid levels in the on-chip reservoirs. Figure 3.16 [106] is a schematic diagram of the microchip electrophoresis system with negative pressure pinched sample injection.

Fang et al. [107] used a high throughput continuous sample introduction interface in NCE. The authors attempted to achieve world-to-chip interfacing by a novel flow-through sampling reservoir featuring a guided overflow design. A cross-channel was designed on a $30 \times 60 \times 3$ mm planar glass

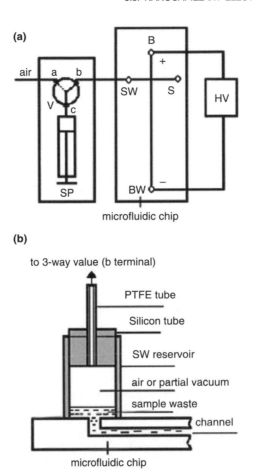

Figure 3.16 Schematic representation of (A) NCE with negative pressure pinched sample injection. SP, syringe pump; V, 3-way valve, HV, high voltage power supply. (B) View of the connection from the three-way valve to the sample waste reservoir SW [106].

chip for NCE. A 20 μL sample reservoir was produced from a section of plastic pipette tip and fixed at one end of the sampling channel. Luo et al. [108] discussed a novel sample injection method in NCE. This injection method used hydrostatic pressure, generated by emptying the sample waste reservoir, for sample loading and electrokinetic force for dispensing. The injection was performed on a double-cross microchip. An advantage of hydrostatic pressure was its ease of generation on a microfluidic chip, without an electrode or external pressure pump, thus allowing sample injection with a minimum number of electrodes. The authors tested the potential of this injection method in a four-separation-channel NCE system. Lacharme and Gijs [109] reported two variants of a new injection technique in NCE, which they called front gate pressure injection and back gate pressure injection. Both enabled a controlled and reproducible sample introduction with reduced bias compared to electrokinetic gated injection.

Automatic sample injection is the new trend in NCE and some workers have described this method of injection. Attiya et al. [110] described designing an automatic sample introduction into NCE. The design contained an interface flow channel with a volume flow resistance of 0.54 to 4.1 × 106 times lower than the volume flow resistance of the electrokinetic fluid required for mixing, reaction, separation, and analysis. The unit was tested with on-chip mixing, reaction, and separation of anti-ovalbumin, and ovalbumin, with good reproducibilities of injection and peak height. Similarly, Lin et al. [111] described automation for continuous analysis in NCE by using flow through sampling. The unit has an autosampler connected online to the microchip and all the steps, that is, sample loading and injection, analysis, and data analyses, were fully automated. The unit was tested using Rhodamin B at different concentrations as the test sample. The fully automatic NCE allowed sequential analysis of a series of samples. Recently, Huang et al. [112] described an integrated NCE with automatic sampling for DNA/RNA amplification. In this unit DNA/RNA samples were first replicated using a micromachine-based polymerase chain reaction module and then transported by a pneumatic micropump to a sample reservoir. The samples were injected electrokinetically into an NCE and separated, followed by their detection by a buried optical fiber. The unit is an important contribution to the fields of molecular biology, genetic analysis, infectious disease detection, and other biomedical applications.

3.3.4. Detectors

The detector in capillary electrophoresis is the main component in nanoanalyses. Many detectors can be used for this purpose but the mass spectrometer is the best one due to its wide ranges and low concentration detection capabilities. In the last few years, time-of-flight–mass spectrometry (TOF–MS) instruments have come onto the market and are available in many sizes, but small instruments are preferred in NCE. Bruker (Billerica, MA) has provided a micro-TOF-MS-LC (2 × 2 × 4 feet) system for nanoanalyses. Bruker also introduced a Q-q-FTMS (Fourier transform mass spectrometer) for proteomics called the APEX-QE. It offers fast, dual quadrupoles, which provides the first stages followed by FTMS for the highest mass accuracy. It can be coupled to NCE and controlled by Bruker's ProteinScape work flow and warehousing software. In addition, NCE-NMR may be useful in nanoseparations. Some reviews and research papers [113–118] have been published on the detection in capillary electrophoresis indicating the mode of hyphenation and working capacities. Brivio et al. [46] described the superamolecular nature of nanoflow electrospray ionization (NESI) time-of-flight (TOF) mass spectrometer (MS) in chip-based microfluid devices. Details on detection in capillary electrophoresis are given in Chapter 4.

3.3.5. Recorder

As in the case of nanoliquid chromatography, the recorder also plays an important role in analyses by printing the electropherograms. All the available NCE instruments are computer based and remote controlled. The data are processed through different software. The most commonly used are Millennium 2000 and ELAN 5000. These are capable of sensing even a single signal and can record it in the form of an electropherogram. The electropherograms can be expanded, integrated, and observed easily. Comparisons of standard and sample electropherograms, nanoscale quantitation, calculations of errors, etc., can be carried out using the software.

3.3.6. Sample Preparation Units

Nanocapillary electrophoresis deals with low amounts of sample and, hence, poor detection is an inherent property of the machine, which calls for further advancements. Among many other proposed ways to avoid this problem, sample preparation and concentration are the most effective tools in NCE. Several workers have attempted to couple sample preparation devices with NCE [119–121]. It has been observed that online hyphenations of sample preparation techniques are useful for analyses of samples having low volume and poor ingredient concentrations. These techniques are highly selective in nature and provide the least interference in analyses due to various impurities present in unknown matrices. The layout and dimensions of an SPE-CE microfluid device is shown in Fig. 3.17 [121] indicating the coupling of SPE with chip-based capillary electrophoresis. The system has been found effective for the analysis of several analytes, including dyes. Recently, Pamme [122] presented a review article on continuous flow separations in microfluid devices. The author described methods of continuous injection, real-time monitoring, as well as continuous collection. Furthermore, the author reported that in continuous flow separation, the sample components are deflected from the main direction of flow, either by means of a force field (electric, magnetic, acoustic, optical, etc.), or by intelligent positioning of obstacles in

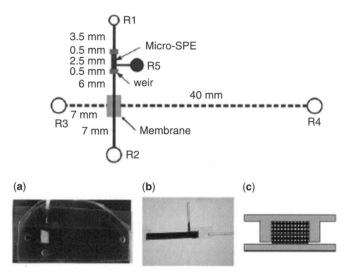

Figure 3.17 Layout and dimensions of the integrated SPE-NCE with (b) photograph of the multilayer device; (c) and (d) micrograph and diagram of the packed micro-SPE column between two shallow weirs [121].

combination with laminar flow profiles. Pamme [122] also reported a large variety of methods with miniaturized versions of larger-scale methods.

Zhang et al. [106] resolved rhodamine123 and fluorescein by using NCE. The migration times, theoretical plate numbers resolution, back flow rate, and peak heights of rhodamine123 and fluorescein are given in Table 3.5.

TABLE 3.5 Effects of Reservoir Levels on Migration Time, Theoretical Plate Numbers, Resolutions, and Peak Heights of Rhodamine123 and Fluorescein in NLC[a,b]

	B		C				F	
A	Rh123	Flu	Rh123	Flu	D	E	Rh123	Flu
0.0	31.4	73.1	1.84	2.14	15.6	0.0000	5.85	11.2
1.7	32.9	80.6	0.94	0.97	12.7	0.0038	9.94	14.8
2.0	33.0	81.1	1.90	2.20	16.0	0.0040	9.90	15.5
3.0	33.4	83.6	1.95	2.29	16.9	0.0051	9.17	13.6
4.0	33.8	86.2	1.97	2.40	17.4	0.0062	8.49	12.5
4.5	33.9	87.0	1.98	2.45	17.8	0.0065	7.75	11.3

[a]Ref. 106.
[b]A, $\Delta h/mm$; B, migration time per second; C, theoretical plate numbers/$10^5 m^{-1}$; D, resolution; E, back flow rate/cm s^{-1}; and F, peak heights per 10^3 μV. Rh123 and Flu are rhodamine123 and fluorescein, respectively.

**TABLE 3.6 NCE Characteristics for Dopamine and Catechol
as a Function of Applied Voltage^{a,b}**

Applied CE Field, V/cm	Migration Time, s	Apparent Mobility, μm·V/s	Peak Area (arbitrary units)	Efficiency, Plates/m
Dopamine				
25	73.0 (±0.5)	55 200 (±490)	1510 (±73)	69 100 (±2300)
50	37.6 (±0.1)	53 700 (±180)	1520 (±66)	93 500 (±1900)
100	20.1 (±0.1)	51 000 (±530)	980 (±110)	98 900 (±1300)
200	10.5 (±0.0)	47 900 (±100)	435 (±15)	54 200 (±2000)
Catechol				
25	124.6 (±1.0)	32 400 (±300)	3470 (±122)	28 000 (±800)
50	65.4 (±0.4)	30 900 (±230)	4050 (±136)	38 900 (±100)
100	36.0 (±0.1)	28 600 (±260)	1170 (±85)	68 000 (±800)
200	18.2 (±0.0)	27 800 (±130)	197 (±5)	100 400 (±5200)

^{a}Ref. 68.
^{b}The numbers shown in parentheses represent absolute standard deviations.

This table indicates quite acceptable migration times, resolution, and values of theoretical plates, making NLC effective and applicable in real samples. The migration times, applied voltages, apparent mobilities, peak areas, and separation efficiencies of dopamine and catechol in NCE are given in Table 3.6. A perusal of this table indicates that migration times for both compounds are quite good, indicating good separation which is decreased as the NCE voltage is increased. Furthermore, the efficiencies of the system were higher at 100 and 200 V/cm applied voltages in dopamine and catechol, respectively [68].

3.4. CONCLUSION

Nowadays, the demand for nanoanalyses is increasing continuously and some advancement has been made in chromatographic and capillary electrophoresis instruments. The integration of all the steps of analysis, that is, sample preparation, injection, separation, and detection on a single chip, is the most difficult task for scientists. In spite of many encouraging advancements in chip LC/CE instrumentation we are still far away from realizing the visions presented a decade ago. Briefly, more advances are needed to turn the dream of real nanochromatography and capillary electrophoresis into a mature analytical tool.

REFERENCES

1. I. Ali, H.Y. Aboul-Enein, *Instrumental methods in metal ions speciation: Chromatography, capillary electrophoresis and electrochemistry*, New York: Taylor & Francis (2006).
2. I. Ali, H.Y. Aboul-Enein, *Chiral pollutants: Distribution, toxicity and analysis by chromatography and capillary electrophoresis*, Chichester: Wiley (2004).
3. P.J. Marríot, P. Haglund, R.C. Ong, *Clin. Chim. Acta*, **328**, 1 (2003).
4. J.C. Giddings, *Multidimensional gas chromatography*, New York: Marcel Dekker (1990).
5. G. Schomburg, *J. Chromatogr. A*, **703**, 309 (1995).
6. I.S. Krull, R.L. Stevenson, K. Mistry, M.E. Swartz, *Capillary electrochromatography and pressurized flow capillary-electrochromatography: An introduction*, New York: HBN Publishing (2000).
7. B. Chankvetadze, *Capillary electrophoresis in chiral analysis*, Chichester: Wiley (1997).
8. L. Szekely, R. Freitag, *Anal. Chim. Acta*, **512**, 39 (2004).
9. J. Xie, J. Shih, Q. He, C. Pang, Y.C. Tui, Y. Miuo, T.D. Leez, *An integrated LC-ESI chip with electrochemical-based gradient generation*, New York: IEEE.
10. Dionex Corporation, USA., http://www.dionex.com.
11. Eksigent, http://www.eksigent.com/hplc/nano.
12. F.Q. Nie, M. Macka, L. Barron, D. Connolly, N. Kentb, B. Paull, *Analyst*, **132**, 417 (2007).
13. A. Manz, N. Graber, H.M. Widmer, *Sensors & Actuators B*, **1**, 244 (1990).
14. J.G. Smits, *Sensors & Actuators A*, **21**, 203 (1990).
15. F.C.M. van de Pol, H.T.G. van Lintel, M. Elwenspoek, J.H.J. Fluitman, *Sensors Actuators A*, **21**, 198 (1990).
16. A. Manz, J.C. Fettinger, E. Verpoorte, S. Haemmerli, H.M. Widmer, *Micro. Sys. Technol.*, 49 (1991).
17. J.C. Fettinger, A. Manz, H. Ludi, H.M. Widmer, *Sensors & Actuators B*, **17**, 19 (1993).
18. Rheodyne, A unit of IDEX Corporation, www.rheodyne.com.
19. JX. Ye, Z. Kuklenyik, L.L. Needham, A.M. Calafat, *Anal. Chem.*, **77**, 5407 (2005).
20. H. Shen, Q. Fang, Z.L. Fang, *Lab. Chip*, **6**, 1387 (2006).
21. E. Chmela, M.T. Blom, H.J. Gardeniers, A. van den Berg, R. Tijssen, *Lab. Chip*, **2**, 235 (2002).
22. Q. He, Integrated nano liquid chromatography system on a chip, Ph.D. Thesis, California Institute of Technology, Pasadena, California (2005).
23. G. Rozing, *LC GC Europe*, **16**, 14 (2003).

24. A. de Mello, *Lab. Chip*, **2**, 48 (2002).

25. C.M. Harris, *Anal. Chem.*, **75**, 64 (2003).

26. G. Ocvirk, E. Verpoorte, A. Manz, M. Grasserbauer, H.M. Widmer, *Anal. Methods Instrumentation*, **2**, 74 (1995).

27. A. Manz, Y. Miyahara, J. Miura, Y. Watanabe, H. Miyagi, K. Sato, *Sensors & Actuators Chem.*, **1**, 249 (1990).

28. B. He, N. Tait, F. Regnier, *Anal. Chem.*, **70**, 3790 (1998).

29. C. Ericson, J. Holm, T. Ericson, S. Hjerten, *Anal. Chem.*, **72**, 81 (2000).

30. J.P. Murrihy, M.C. Breadmore, A.M. Tan, M. McEnery, J. Alderman, C. O'Mathuna, A.P. O'Neill, P. O'Brien, N. Advoldvic, P.R. Haddad, J.D. Glennon, *J. Chromatogr. A*, **924**, 233 (2001).

31. K. Killeen, H. Yin, D. Sobek, R. Brennen, T.V.D. Goor, Chip-LC/MS: HPLC-MS using polymer microfluidics. *Presented at Seventh International Conference on Micro Total Analysis Systems (microTAS 2003)*. Squaw Valley, California, October 5–9, 481 (2003).

32. J. Xie, J. Shih, Y. Miao, T.D. Lee, Y.-C. Tai, Complete gradient-LC- ESI system on a chip for protein analysis. *Proceedings of the 18th IEEE International Conference on Micro Electro Mechanical Systems (MEMS 2005)*. Miami Beach, Florida, Jan 30–Feb 2, 778 (2005).

33. Q. He, C. Pang, Y.-C. Tai, T.D. Lee, Ion liquid chromatography on-a- chip with beads-packed parylene column. *Proceedings of the 17th IEEE International Conference on Micro Electro Mechanical Systems (MEMS 2004)*. Maastricht, The Netherlands, Jan 25–29 (2004).

34. G.A. Lord, D.B. Gordon, P. Myers, B.W. King, *J. Chromatogr. A*, **768**, 9 (1997).

35. E.S. Yeng, W.G. Kuhr, *Anal. Chem*, **63**, 275 (1991).

36. M.W.F. Nielen, *J. Chromatogr.*, **588**, 321 (1992).

37. T. Wang, R.A. Hartwick, *J. Chromatogr.*, **607**, 119 (1992).

38. H. Poppe, *Anal Chem.*, **64**, 1908 (1992).

39. J.L. Beckers, *J. Chromatogr. A*, **679**, 153 (1994).

40. W.T. Kok, *Chromatographia*, **51**, 52 (2000).

41. A.M. Leach, M. Heisterkamp, F.C. Adams, G.M. Hieftje, *J. Anal. Atom. Spectrom.* **15**, 151 (2000).

42. M.J. Mills, J. Maltas, W.J. Lough, *J. Chromatogr. A*, **759**, 1 (1997).

43. A. Cappiello, G. Famiglini, P. Palma, F. Mangani, *Anal. Chem.*, **74**, 3547 (2002).

44. J. Masuda, D.M. Maynard, M. Nishimura, T. Ueda, *J. Chromatogr. A*, **1063**, 57 (2005).

45. A. Zamfir, Z. Vukelić, L. Bindila, J. Peter-Katalinić, R. Almeida, A. Sterling, M. Allen, *J. Am. Soc. Mass Spectrom.*, **11**, 1649 (2004).

46. M. Brivio, R.E. Oosterbroek, W. Verboom, A. van den Berg, D.N. Reinhoudt, *Lab. Chip*, **5**, 1111 (2005).

47. C.Y. Shih, Y. Chen, J. Xie, Q. He, Y.C. Tai, *J. Chromatogr. A*, **1111**, 272 (2006).
48. J.L. Veuthey, S. Souverain, S. Rudaz, *Ther. Drug Monit.*, **26**, 161 (2004).
49. F. Pragst, *Anal. Bioanal. Chem.*, **388**, 1393 (2007).
50. M. Tokeshi, T. Minagawa, T. Kitamori, *Anal. Chem.*, **72**, 1711 (2000).
51. H. Hisamoto, T. Horiuchi, M. Tokeshi, A. Hibara, T. Kitamori, *Anal. Chem.*, **73**, 1382 (2001).
52. H. Hisamoto, T. Horiuchi, K. Uchiyama, M. Tokeshi, A. Hibara, T. Kitamori, *Anal. Chem.*, **73**, 5551 (2001).
53. T. Minagawa, M. Tokeshi, T. Kitamori, *Lab. Chip*, **1**, 72 (2001).
54. K. Vanhoutte, W. van Dongen, I. Hoes, F. Lemiere, E.L. Esmans, H.V. Onckelen, E. van den Eeckhout, R.E.J. van Soest, A.J. Hudson, *Anal. Chem.*, **69**, 3161 (1997).
55. A.R. de Boer, B. Bruyneel, J.G. Krabbe, H. Lingeman, W.M.A. Niessen, H. Irth, *Lab. Chip*, **5**, 1286 (2005).
56. A. Manz, D.J. Harrison, E.M.J. Verpoorte, J.C. Fettinger, A. Paulus, H. Ludi, M. Winder, *J. Chromatogr. A.*, **593**, 253 (1992).
57. A. Manz, H. Becker (Eds.), *Microsystem technology in chemistry and life sciences*, Berlin, Springer-Verlag, (1992).
58. C.K. Fredrickson, Z.H. Fan, *Lab. Chip*, **4**, 526 (2004).
59. W.C. Sung, H. Makamba, S.H. Chen, *Electrophoresis*, **26**, 1783 (2005).
60. I. Nischang, U. Tallarek, *Electrophoresis*, **28**, 611 (2007).
61. A.D. Zamfir, *J. Chromatogr. A.*, **1159**, 2 (2007).
62. L. Zhang, X.F. Yin, *J. Chromatogr. A*, **1137**, 243 (2006).
63. A. Manz, K.D. Luckas, *Science*, **222**, 266 (1983).
64. M.V. Bazylenko, M. Gross, *J. Vac. Sci. Technol. A*, **14**, 2994 (1996).
65. K.J. An, D.H. Lee, J.B. Yoo, J. Lee, G.Y. Yeom, *J. Vac. Sci. Technol. A*, **17**, 1483 (1999).
66. H. Nakanishi, H. Abe, T. Nishimoto, A. Arai, *Bunseki Kagaku*, **47**, 361 (1998).
67. T. Ujji, T. Kikuchi, T. Ichiki, Y. Horiike, *J. Appl. Phys.*, **39**, 3677 (2000).
68. D.J. Jackson, J.F. Naber, T.J. Roussel, Jr. M.M. Crain, K.M. Walsh, R.S. Keynton, R.P. Baldwin, *Anal. Chem.*, **75**, 3643 (2003).
69. G.J.M. Bruin, *Electrophoresis*, **21**, 3931 (2000).
70. N.A. Lacher, K.E. Garrison, R.S. Martin, S.M. Lunte, *Electrophoresis*, **22**, 2526 (2001).
71. D.R. Reyes, D. Iossifidis, P.-A. Auroux, A. Manz, *Anal. Chem.*, **74**, 2623 (2002).
72. P.A. Auroux, D. Iossifidis, D.R. Reyes, A. Manz, *Anal. Chem.*, **74**, 2637 (2002).
73. P.A. Greenwood, G.M. Greenway, *TrAC*, **21**, 726 (2002).
74. M. Urbanek, L. Krivankova, P. Bocek, *Electrophoresis*, **24**, 46 (2003).

75. T.B. Stachowiak, F. Svec, J.M.J. Fréchet, *J. Chromatogr.*, **1044**, 97 (2004).

76. C.H. Lin, T. Kaneta, *Electrophoresis*, **25**, 4058 (2004).

77. H.J. Issaq, G.M. Janini, K.C. Chan, T.D. Veenstra, *J. Chromatogr. A*, **1053**, 37 (2004).

78. T. Vilkner, D. Janasek, A. Manz, *Anal. Chem.*, **76**, 3373 (2004).

79. C.H. Xie, H.J. Fu, J.W. Hu, H.F. Zou, *Electrophoresis*, **25**, 4095 (2004).

80. W. Li, D.P. Fries, A. Malik, *J. Chromatogr. A*, **1044**, 23 (2004).

81. C.W. Klampfl, *J. Chromatogr. A*, **1044**, 131 (2004).

82. F. Svec, *J. Separation Sci.*, **28**, 729 (2005).

83. E. Klodzinska, D. Moravcova, P. Jandera, B. Buszewski, *J. Chromatogr. A*, **1109**, 51 (2006).

84. P.S. Waggoner, H.G. Craighead, *Lab. Chip*, **7**, 1238 (2007).

85. L.M. Fu, R.J. Yang, G.B. Lee, Y.J. Pan, *Electrophoresis*, **24**, 3026 (2003).

86. C.C. Chang, R.J. Yang, *Microfluid Nanofluid*, **3**, 501 (2007).

87. C. Futterer, N. Minc, V. Bormuth, J.H. Codarbox, P. Laval, J. Rossier, J.L. Viovy, *Lab. Chip*, **4**, 351 (2004).

88. S.C. Jacobson, R. Hergenroder, L.B Kounty, J.M. Ramsey, *Anal. Chem.*, **66**, 1114 (1994).

89. D.J. Harrison, K. Fluri, K. Seiler, Z. Fan, C.S. Effenhauser, A. Manz, *Science*, **261**, 895 (1993).

90. J.P. Alarie, S.C. Jacobsson, C.T. Cultberson, J.M. Ramsey, *Electrophoresis*, **21**, 100 (2000).

91. L.L. Shultz-Lockyear, C.L. Colyer, Z.H. Fan, D.J. Harrison, *Electrophoresis*, **20**, 529 (1999).

92. S.C. Jacobson, S.V. Ermakov, J.M. Ramsey, *Anal. Chem.*, **71**, 3273 (1999).

93. C.X. Zhang, A. Manz, *Anal. Chem.*, **73**, 2656 (2001).

94. C.D. Thomas, S.C. Jacobson, J.M. Ramsey, *Anal. Chem.*, **76** 6053 (2004).

95. J.P. Alarie, S.C. Jacobson, J.M. Ramsey, *Electrophoresis*, **22**, 312 (2001).

96. B.E. Slentz, N.A. Penner, F. Regnier, *Anal. Chem.*, **74**, 4835 (2002).

97. Q. Fang, F.R. Wang, S.L. Wang, S.S. Liu, S.K. Xu, Z.L. Fang, *Anal. Chim. Acta*, **390**, 27 (1999).

98. X. Bai, H.J. Lee, J.S. Rossier, F. Reymond, H. Schafer, M. Wossner, H.H. Girault, *Lab. Chip*, **2**, 45 (2002).

99. D. Solignac, M.A.M. Gijs, *Anal. Chem.*, **75**, 1652 (2003).

100. D. Kaniansky, M. Masár, M. Danková, R. Bodor, R. Rákocyova, M. Pilna, M. Jöhnk, B. Stanislawski, S. Kajan, *J. Chromatogr. A*, **1051**, 33 (2004).

101. S.J. Baldock, P.R. Fielden, N.J. Goddard, H.R. Kretschmer, J.E. Prest, B.J. Treves Brown, *J. Chromatogr. A*, **1042**, 181 (2004).

102. H. Gai, L. Yu, Z. Dai, Y. Ma, B.C. Lin, *Electrophoresis*, **25**, 1888 (2004).

103. Z. Wu, H. Jensen, J. Gamby, X. Bai, H.H. Girault, *Lab. Chip*, **4**, 512 (2004).

104. S.I. Cho, S.H. Lee, D.S. Chung, Y.-K. Kim, *J. Chromatogr., A*, **1063**, 253 (2005).

105. S. Buttgenbach, R. Wilke, *Anal. Bioanal. Chem.*, **383**, 733 (2005).

106. L. Zhang, X. Yin, Z. Fang, *Lab. Chip*, **6**, 258 (2006).

107. Q. Fang, G.M. Xu, Z.L. Fang, *Anal. Chem.*, **74**, 1223 (2002).

108. Y. Luo, D. Wu, S. Zeng, H. Gai, Z. Long, Z. Shen, Z. Dai, J. Qin, B. Lin, *Anal. Chem.*, **78**, 6074 (2006).

109. F. Lacharme, M.A. Gijs, *Electrophoresis*, **27**, 2924 (2006).

110. S. Attiya, A.B. Jemere, T. Tang, G. Fitzpatrick, K. Seiler, N. Chiem, D.J. Harrison, *Electrophoresis*, **22**, 318 (2001).

111. C.C. Lin, G.B. Lee, S.H. Chen, *Electrophoresis*, **23**, 3550 (2002).

112. F.C. Huang, C.S. Liao, G.B. Lee, *Electrophoresis*, **27**, 3297 (2006).

113. A. von Brocke, G. Nicholson, E. Bayer, *Electrophoresis*, **22**, 1251 (2001).

114. S. Buttgenbach, M. Michalzik, R. Wilke, *Eng. Life Sci.*, **6**, 449 (2006).

115. B.H. Huynh, B.A. Fogarty, R.S. Martin, S.M. Lunte, *Anal. Chem.*, **76**, 6440 (2004).

116. F. Kitagawa, T. Tsuneka, Y. Akimotot, K. Sueyoshi, K. Uchiyama, A. Hattori, K. Otsuka, *J. Chromatogr. A*, **1106**, 36 (2006).

117. K. Shik-Sin, Y.H. Kim, J.A. Min, S.M. Kwak, S.K. Kim, E.G. Yang, J.H. Park, B.K. Ju, T.S. Kim, J.Y. Kang, *Anal. Chim. Acta.*, **573**, 164 (2006).

118. B.H. Huynh, B.A. Fogarty, P. Nandi, S.M. Lunte, *J. Pharm. Biomed. Anal.*, **42**, 529 (2006).

119. J.D. Ramsey, G.E. Collins, *Anal. Chem.*, **77**, 6664 (2005).

120. S. Tumikoski, N. Vikkala, S. Rovio, A. Hokkanen, H. Siren, S. Franssila, *J. Chromatogr. A*, **1111**, 258 (2006).

121. Z. Long, Z. Shen, D. Wu, J. Qin, B. Lin, *Lab. Chip*, **7**, 1819 (2007).

122. N. Pamme, *Lab. Chip*, **7**, 1644 (2007).

CHAPTER 4

DETECTION IN NANOCHROMATOGRAPHY AND NANOCAPILLARY ELECTROPHORESIS

4.1. INTRODUCTION

Basically, the goal of analytical academicians, scientists, and clinicians is to achieve nano or low detection limits in biological and environmental matrices, which is a formidable analytical challenge. To achieve this task the role of the detector is very important in chromatographic and capillary electrophoresis paraphernalia. It is known that the sensitivity of analytes increases with decreasing column internal diameter due to low flow rate, which causes a reduction of chromatographic dilution. The chromatographic dilution factor is proportional to the square of the radii and to the square root of the length and plate height of the column. However, in nanochromatography and nano-capillary electrophoresis relatively low sample volumes are injected (20 to 60 nL) resulting in poor detection and sensitivity, which may be considered a drawback of NLC and NCE. In order to overcome this drawback, the use of appropriate detectors is a very important issue. Various detectors have been used in microfluidic chromatography and capillary electrophoresis. Mostly mass spectrometer based detectors have been used in NLC and NCE. However, detectors such as fluorescence, electrochemical, element specific, and others have also been used in these modalities of chromatography and capillary electrophoresis. The state of the art detectors in microfluidic devices is discussed in this chapter.

Nanochromatography and Nanocapillary Electrophoresis. By Ali, Aboul-Enein, and Gupta
Copyright © 2009 John Wiley & Sons, Inc.

4.2. MASS SPECTROMETER DETECTORS

Of course, mass spectrometry (MS) is gaining importance as the method of detection in nanochromatography and nanocapillary electrophoresis for various applications of biological and environmental interest, because it can detect any compound at nano or low levels of detection. MS detectors are useful to achieve improved sensitivity, structural information, unequivocal identification of compounds, and co-migration of analytes. Moreover, hyphenation of these detectors with NLC and NCE, through different nanospray interfaces, is most effective, with reproducible results. NLC-MS and NCE-MS coupling are highly compatible due to low flow rates (50 to 600 nL/min) [1–4]. But complete characterization of real samples using mass detectors is a complex separation task due to the complexity of interfacing sample separation and mass detection. Tedious sample preparation procedures, including separation and desalting, are required for complex samples before mass detection. Microchips have been integrated to facilitate sample preparation, sample introduction, and mass detection.

Some reviews [5–7] have appeared on NCE-electrospray ionization-mass spectrometry (NCE-ESI-MS) discussing various factors responsible for detection. Recently, Zamfir [8] reviewed sheathless interfacing in NCE-ESI-MS in which the authors discussed several issues related to sheathless interfaces. Feustel et al. [9] attempted to couple mass spectrometry with microfluidic devices in 1994. Other developments in mass spectroscopy have been made by different workers. McGruer and Karger [10] successfully interfaced a microchip with an electrospray mass spectrometer and achieved detection limits lower than 6×10^{-8} mole for myoglobin. Ramsey and Ramsey [11] developed electrospray from small channels etched on glass planar substrates and tested its successful application in an ion trap mass spectrometer for tetrabutylammonium iodide as model compound. Desai et al. [12] reported an electrospray microdevice with an integrated particle filter on silicon nitride.

Zhang et al. [13] developed an NCE-ESI-MS set up for analysis at nanolevels (Fig. 4.1). This figure indicates coupling of NCE with ESI-MS by means of a liquid junction. The connection of the exit port of the chip to the electrospray is an important issue and many different designs have been developed for this purpose [12,14–19]. The separation channel is connected to an ESI capillary via a liquid junction. A capillary is attached to the microdevice as an extension of the separation channel, which allows effluent to be switched either to the mass spectrometer or an online UV detector. In this arrangement (integration of a microfabricated pneumatic nebulizer at the electrospray exit port) the dead volume may be reduced, resulting in an improvement in separation and stability of the electrospray [13]. Licklider et al. [20] described a

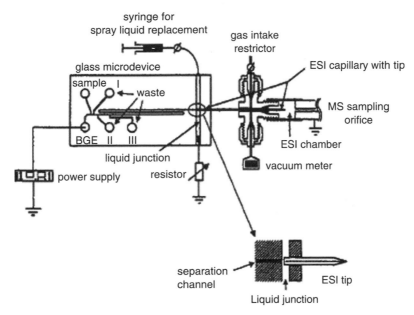

Figure 4.1   A schematic diagram of a microdevice with subatmospheric electrospray interface. The expanded view shows the coupling of the ESI tip with the separation channel in the liquid junction [13].

micromachining process for fabricating a mass spectrometry electrospray source on a silicon chip. The authors reported their work as an efficient electrospray interface to mass spectrometry, useful for detection of various molecules at the nanolevel.

Kameoka et al. [21] developed and used an electrospray ionization system incorporating a triangular thin film polymer tip bonded to a microfluidic channel. Brivio et al. [22] integrated a microfluidic chip for biochemical reactions into a sample plate for matrix-assisted laser desorption/ionization–time-of-flight–mass spectrometry (MALDI-TOF-MS). The effectiveness of the system has been successfully demonstrated with several examples of (bio)-chemical reactions. Le Gac et al. [23] designed a nib-shaped microelectrospray emitter in microfluidic devices. The micronib sources were successfully tested on an LCQ Deca XP¹ ion trap mass spectrometer by using peptide samples at varying concentrations. Huikko et al. [24] developed a coupling of a polydimethylsiloxane (PDMS) microchip with ESI-MS. The hydrophobic PDMS aided the formation of a small Taylor cone in the ESI process and facilitated straightforward and low cost batch production of ESI-MS chips. The authors examined the performance of PDMS devices in the ESI-MS by analyses of some pharmaceutical compounds and amino acids. Lazar et al. [25] described a microfabricated device integrating a monolithic polymeric separation channel, an injector, and an interface for ESI-MS detection. The microchip ESI interface as a semipermeable gate and on-chip junction between the separation channel and ESI emitter is shown in Fig. 4.2. The authors reported the ESI interface allowed hours of stable operation at flow rates generated by the monolithic column. The dimensions of one processing line were sufficiently small to enable the integration of four- to eight-channel multiplexed structures on a single substrate. These workers tested the unit by analyzing protein samples with detection at femtomole levels. Zhang and coworkers [26,27] developed a layout of a multiple ESI chip, which has been commercialized for real-world sample bioanalysis. A channel was etched across the silicon plate and a nozzle (10 mm i.d., 630 mm o.d.) was connected beyond the end of the channel, which is shown in Fig. 4.3. The reservoir etched in the back of the plate could hold the analyte for direct ionization.

Kauppila et al. [28] developed a microfabricated heated nebulizer chip for atmospheric pressure photoionization-mass spectrometry. Various materials have been used to design and develop hyphenation of microfluidic devices and ESI-MS. These materials are photoresist SU-8 [29,30], polymers [31,32], and glassy carbon [33]. Thorslund et al. [34] developed a chip on which sample injection, separation, and ESI-emitter structures are integrated

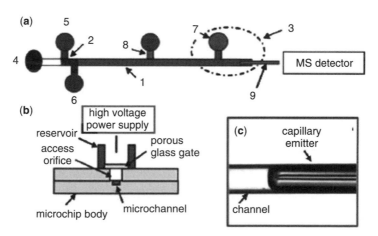

Figure 4.2 A schematic diagram of an integrated polymer monolith NCE with ESI-MS detection. (a) 1, separation channel; 2, double-T injector; 3, ESI source; 4, eluent reservoir; 5, sample inlet reservoir; 6, sample waste reservoir; 7, eluent waste reservoir that houses a porous glass gate; 8, side channel for flushing the monolithic channel; and 9, ESI emitter. (b) Cross-sectional view of reservoir 7, showing the position of the semipermeable glass gate. (c) Image of on-chip junction between the separation channel and the ESI emitter [25].

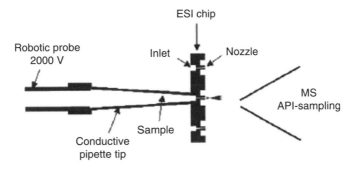

Figure 4.3   Diagram of the layout of a multiple ESI chip [27].

in a single platform, eliminating problems at module connections. Yang et al. [35] integrated a cycloolefin polymer-based chip with a sheathless capillary tip for ESI-MS. Yue et al. [36] carried out a study on glass membrane prepared for conducting high voltage for generation of liquid electrospray. Brivio et al. [37] developed two interfaces, a monolithic design (ionization assisted by on-chip gas nebulization) and a modular approach (commercially available tips), enabling hyphenation of nanoflow ESI time-of-flight MS with microfluidic devices. Xie et al. [38] integrated an NLC chip to analyze a complex mixture of proteins. They used a low volume static mixer, a column packed with silica-based reversed-phase support, integrated frits for bead capture, and an electrospray nozzle.

Basically, NCE-ESI-MS coupling via ESI ionization is considered one of the most important innovations in genomics [39], proteomics [40], drug analysis [41], glycomics [42], and biomarker discovery [43] research. Also, because of the increase in heat transfer observable in miniaturization, the use of higher voltages is possible, thus providing faster separations compared to larger systems [44]. The make up liquid is not a necessary component and a sheath flow interface is difficult to build up in NCE. Therefore, the interface is almost based on a sheathless design containing a nanospray emitter. Normally, the emitter is built up on a separation channel or a needle embedded onto the chip via holders and joints. Some workers [5,6,45] have discussed merits and demerits of the NCE/ESI-MS technique but its merits are more important in several analytical fields. Akashi et al. [46] developed an NCE/ESI-MS design with an external emitter, which is shown in Fig. 4.4. Basically, this is a refinement of the design of Tachibana et al. [47]. This figure indicates a quartz microchip (68.5 × 12.5 × 6.05 mm) with crossed-type microchannels (82 × 36 μm). An electrospray needle 27 mm long is attached with a polyetheretherketone (PEEK) screw and a sleeve through a guide hole with an i.d. of 370 μm at the end of the separation microchannel. This set up was

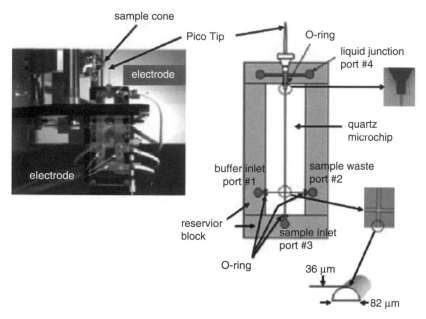

Figure 4.4   NCE set up with a spray needle for ESI [8].

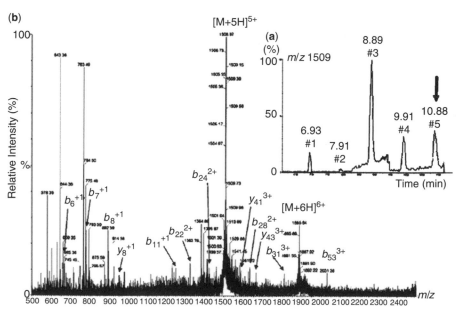

Figure 4.5   NCE/ESI QTOF-MS and tandem MS electropherograms of DNA-binding domain of human TTAGGG repeat binding factor 2 (hTRF2 DBD) protein. (A) Mass electropherogram of hTRF2 DBD $[M+5H]^{5+}$ ion at m/z 1509, and (B) top-down by CID MS/MS of ion corresponding to the peak number 5 at 10.88 min [8].

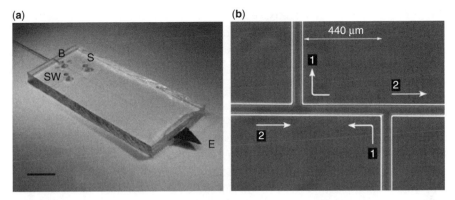

Figure 4.6 PDMS microchip. (a) PDMS microchip with (E) a graphite-coated emitter tip, (S) reservoirs for sample, (SW) sample waste and (B) buffer. (b) Double-T injection cross of the PDMS-device: (1) channel path and (2) buffer flow [8].

used successfully for various applications; one example is shown in Fig. 4.5: the separation, identification, ionization, and sequencing of basic proteins. Sometimes, the connections of an emitter introduce dead volumes between the microchip channel and capillary. Therefore, Lazar et al. [6] and Sung et al. [5] attempted to overcome this limitation based on emitter on chip etching using PDMS as the microchip substrate. Basically, the work of Lazar et al. [6] and Sung et al. [5] is based on the same approaches reported earlier in the literature [30,48,49]. Svedberg et al. [50] developed PDMS microchips using a nickel mold replicated from a dry etched silicon wafer. The tip for ESI-MS had a groove with parallel walls. Bergquist and coworkers [34,51] designed two advanced sheathless ESI microchips containing sample injection, capillary electrophoretic separation and electrospray emitter (Fig. 4.6). This chip contains sample reservoirs, wastes, and buffers. A tapered fused silica capillary was inserted into the buffer reservoir and sealed with a droplet of PDMS. Hop et al. [52] described a chip-based nanoelectrospray ionization source to compare the ionization efficiencies of compounds with different physicochemical properties. The data indicated that the ionization efficiencies varied less with the chip-based device than with LC/MS.

4.3. FLUORESCENCE DETECTORS

Of course, fluorescence detectors are very useful tools to achieve nano detection tasks but have certain limitations, which depend on the characteristic features of compounds. This sort of detector has been used in nano-LC and CE with few applications [53,54]. The interesting features of fluorescence detectors are their high selectivity and sensitivity than UV-visible detectors. The major drawback associated with these detectors is the restriction to only fluorescent compounds and sometimes they require pre-derivatization of non-fluorescent molecules. As an alternative method to fluorescence detection is chemiluminescence detection, which is characterized by a simple and inexpensive optical system that requires no light source, providing low background with excellent sensitivity. Several chemiluminescence detectors have been reported and used in NCE, including luminol [55], peroxyoxalate [56], and Ru(bpy)$_3^{3+}$ [57]. Shin et al. [58] described a miniaturized fluorescence detection chip for NCE immunoassay of atrazine. The photodiode with fluorescence filter was embedded in a PDMS microfluidic chip and placed just below the microfluidic channel. The microchip is shown in Fig. 4.7. The microchip is 25 mm wide and 50 mm long, with a four-way injection, 27 mm long separation channel, and 8 mm long sample loading channels. Schulze et al. [59] developed a two photon excited (TPE) fluorescence detector for NCE. The authors evaluated this technique as an alternative to common photon excitation in the deep UV spectral range. They reported TPE enabled fluorescence detection of unlabeled aromatic compounds, even in non-deep UV-transparent

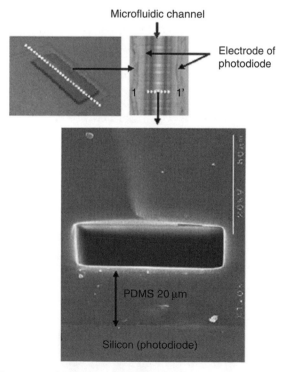

Figure 4.7 Microscope images and a cross-sectional SEM image of the miniaturized fluorescence detection chip (70 × 12 μm) [58].

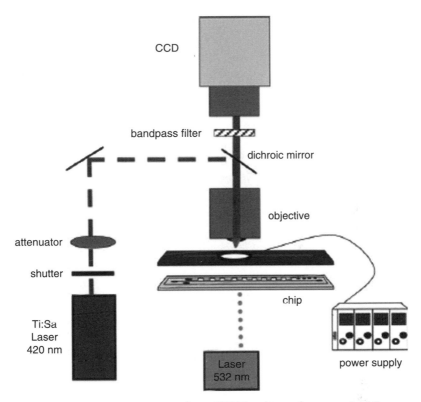

Figure 4.8   A schematic representation of NCE and two photon excited fluorescence detection. Excitation light: dashed line. Emission light: solid line. The dotted line displays light for visualization [59].

microfluidic chips. A schematic drawing of the system is shown in Fig. 4.8, indicating a coupling of microchip electrophoresis and two-photon excited fluorescence detection. Jorgensen et al. [60] designed a chip-based device for measuring chemiluminescence. Jackson et al. [61] designed a miniaturized, battery-powered device with high voltage power supply, electrochemical detection, and interface circuits for NCE. The interface circuit also maintains the detection reservoir at ground potential and allows channel currents to be measured likewise. The device was tested by performing separations of dopamine and catechol and by laser-induced fluorescence visualization.

4.4. ELECTROCHEMICAL DETECTORS

Electrochemical detectors are also useful devices in NLC and NCE due to their increase in sensitivity. But, unfortunately, these have not found wide use in the nano world because of their one limitation, of sensitivity towards electrochemically active compounds only. Besides, electrode alignment within the capillary is necessary for adequate and reproducible measurements, which is a tedious job [62–64]. Wilke and Buttgenbach [65] described an NCE with fully integrated electrodes for amperometric detection. The unit can be used for analyses with no extra mechanical equipment for electrode insertion. Figure 4.9 shows a buffer waste reservoir, with two electrodes for electrochemical detection on the left and a high voltage electrode on the right with a separation channel to the left. The device was used successfully for the analyses of hydrogen peroxide, ascorbic acid, and uric acid simultaneously. Liao et al. [66] reported a poly-(methyl methacrylate) (PMMA)-based microchip, with an integrated gold nanoelectrode ensemble (GNEE) and a quartet-T loading channel for electrochemical detection. The authors fabricated the GNEE electrode using an electrodeless deposition process on a thin polycarbonate film and bonded directly onto a PMMA substrate. Hsueh et al. [67] and Arora et al. [68] fabricated electrochemical cells in silicon and PMMA, respectively, for detection of electrochemiluminescent compounds.

Guijt et al. [69] reported four-electrode capacitively coupled conductivity detection in NCE. The glass microchip consisted of a 6 cm etched channel (20×70 μm cross-section) with silicon nitride covered walls. Laugere et al. [70] described chip-based, contactless four-electrode conductivity detection in NCE. A 6 cm long, 70 μm wide, and 20 μm deep channel was etched on a glass substrate. Experimental results confirmed the improved characteristics of the four-electrode configuration over the classical two-electrode detection set up. Jiang et al. [71] reported a mini-electrochemical detector in NCE,

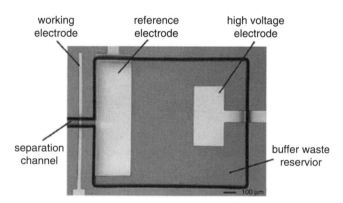

Figure 4.9 The buffer waste reservoir with two electrodes for electrochemical detection on the left and a high voltage electrode on the right. The separation channel extends to the left [65].

that is, of $3.6 \times 5.0 \, cm^2$, width and length, respectively. The detection performance of the new system was demonstrated by the detection of epinephrine using an integrated PDMS/glass microchip.

4.5. ELEMENT SPECIFIC DETECTORS

Generally, the separation channel of microfluidic devices is shorter than a normal CE capillary resulting in shorter analysis time (in seconds). Therefore, coupling of NCE to inductively coupled plasma (ICP) improves overall sample throughput. Really, NCE interfacing to ICP is a challenging task due to the smaller sample size, which requires efficient nebulization and transport of NCE eluent to ICP. However, Song et al. [72,73] reported an interface of NCE with ICP. Typical NCE-ICP interfaces are shown in Fig. 4.10, indicating the pattern of hyphenation. The dimensions of the spray chamber are important for the peak shape and size as large chambers

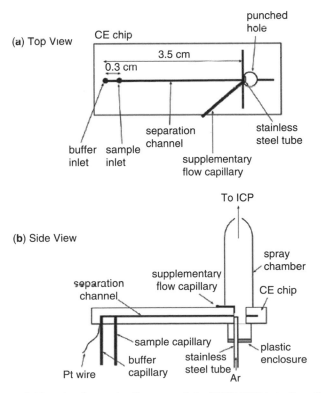

Figure 4.10   A schematic diagram of the NCE-ICP interface [73].

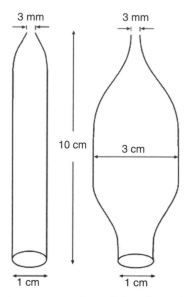

Figure 4.11 A schematic representation of the dimensions and shapes of spray chambers [75].

may broaden the NCE peak. Two types of chambers are used (Fig. 4.11), that is, smaller and larger. Good results were obtained with the small chamber. Li et al. [74] described direct interfacing of NCE to a sensitive and selective detector for atomic fluorescence spectrometry for rapid speciation analysis. Hui et al. [75] developed and described an interface of NCE to inductively coupled plasma atomic emission spectrometry (NCE-ICP-AES), based on

TABLE 4.1   Experimental Conditions of the Chip-Based CE-ICP System[a]

ICP forward power	1.15 kW
Frequency of rf power supply	27 MHz
Number of simultaneous channels	12
Observation height	10 mm above the load coil
Plasma Ar flow rate	12 L min^{-1}
Auxiliary Ar flow rate	1.0 L min^{-1}
Carrier Ar flow rate	1.0 L min^{-1}
Supplementary buffer flow rate	3.5 μL min^{-1}
Emission lines	Ba II 455.4 nm
	Mg I 285.2 nm

[a]Ref. 75.

cross-flow nebulization. A PDMS NCE chip with conventional cross-channel layout was used. The nebulization and transport efficiency of the NCE-ICP-AES interface was approximately 10%. Barium and magnesium ions were eluted from NCE within 30 seconds. The experimental conditions of the NCE-ICP system are given in Table 4.1, which indicates various working variables.

4.6. MISCELLANEOUS DETECTORS

In addition to the detectors discussed above, other devices, such as UV-visible, conductivity and thermal lens microscope (TLM) have been coupled with NLC and NCE and used for detection at nano or low levels. As in conventional LC and CE [76–78], UV-visible detectors are not very common in nanochromatography and nanocapillary electrophoresis due to their low detection limits and need for slightly high mobile phase volume. However, on-column detection for accurate qualitative and quantitative analyses is possible using these detectors. Moreover, the coupling of these detectors with chromatography and capillary electrophoresis is quite good under varied experimental conditions. But, sometimes, short path-lengths of the capillaries in LC and CE create a problem with detection at the nano level. Some workers attempted to overcome this limitation by using Z-cells in detectors [79,80]. Uchiyama et al. [81] described a connection between a capillary electrophoresis chip and a TLM. Later on, Kitagawa et al. [82] described a coupling of a TLM detector with NCE using an interface chip (IFChip) (Fig. 4.12) for achieving sensitive detection with high reproducibility. The authors reported 3,900,000-fold increase in the sensitivity. Laugere et al. [83] developed an electronic interface to detect liquid conductivity in NCE by capacitive four-electrode coupling.

Mogensen et al. [84] described the fabrication and working of a silicon-based NCE chip with integrated optical waveguides for absorption detection. A 750 µm long U-shaped detection cell was used to augment detection. Ro et al. [85] discussed an integrated light collimating system for extended optical-path-length absorbance detection in NCE. The collimating system was made of the same material as the chip (PDMS), and it is integrated into the chip during the molding of the NCE microchannels. To block stray light, two rectangular apertures, realized by a specially designed three-dimensional microchannel, were made on each end of the detection cell. The authors reported the detection sensitivity was increased by 10 times in the newly developed absorbance detection cell. Ma et al. [86] described a hybrid microdevice with a thin PDMS membrane on the detection window for UV absorbance

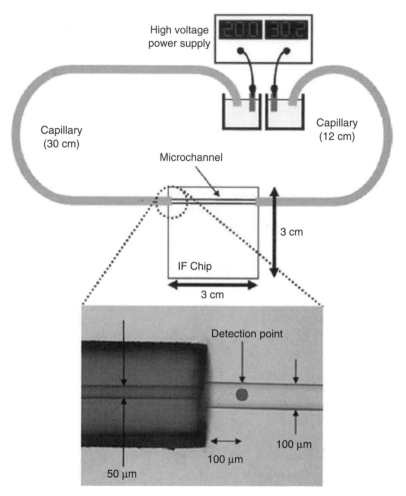

Figure 4.12   A schematic diagram of NCE-IF Chip-TLM [82].

detection. The thickness of the PDMS membrane was about $100\,\mu m$, with high UV transmittance. Li et al. [87] described theoretical study of a contact-less conductivity detector in NCE, based on the construction of an equivalent circuit mode of the high frequency conductivity detector and the design of the structure. The detector has many advantages, such as wide application range, simple structure, and easy integration. McBrady and Synovec [88] described a microfabricated refractive index gradient (μ-RIG) chip-based detector in reversed phase NLC with mobile phase gradient elution. The authors reported that viscosity change during the mobile phase gradient separation was found to shift the on-chip merge position of the detected concentration gradient.

4.7. CONCLUSION

Mostly, mass spectrometers have been used widely for nanodetection in NLC and NCE due to its low detection limits and ease of hyphenation with microfluidic devices. However, attempts have been made to couple other detectors with NLC and NCE. The state of the art of hyphenation of detectors in NLC and NCE is still in its development stage. More advances are expected in the near future for detection at extremely low concentrations for a wide range of molecules.

REFERENCES

1. E. Edwards, J. Thomas-Oates, *Analyst*, **130**, 13 (2005).
2. J. Abian, A.J. Oosterkamp, E. Gelpi, *J. Mass. Spectrom.*, **34**, 244 (1999).
3. V. Tomer, *Chem. Rev.*, **101**, 297 (2001).
4. A.D. Zamfir, J. Peter-Katalinic, *Electrophoresis*, **25**, 1949 (2004).
5. W.C. Sung, H. Makamba, S.H. Chen, *Electrophoresis*, **26**, 1783 (2005).
6. I.M. Lazar, J. Grym, F. Foret, *Mass Spectrom. Rev.*, **25**, 573 (2006).
7. T.B. Stachowiak, F. Svec, J.M.J. Frechet, *J. Chromatogr. A*, **1044**, 97 (2004).
8. A.D. Zamfir, *J. Chromatogr. A*, **1159**, 2 (2007).
9. A. Feustel, J. Muller, V. Rclling, In *Proceedings of Micro Total Analysis Systems 1994*, Dordrecht, The Netherlands: Kluwer Academic Publishers (1994).
10. N.E. McGruer, B.L. Karger, *Anal. Chem.*, **69**, 426 (1997).
11. R.S. Ramsey, J.M. Ramsey, *Anal. Chem.*, **69**, 1174 (1997).
12. A. Desai, Y.C. Tai, M.T. Davis, T.D. Lee, *International Conference on Solid State Sensors and Actuators (Transducers-97)*, Chicago, IL, June 16–19, p. 927 (1977).
13. B. Zhang, F. Foret, B.L. Karger, *Anal. Chem.*, **72**, 1015 (2000).
14. H. Lui, F. Foret, C. Felten, B. Zhang, P. Jedrzejewski, B.L. Karger, *Proc. 46th ASMS Conf. Mass Spectrom. Allied Topics*, Orlando, FL (1998).
15. J. Li, P. Thibault, N. Bing, C.D. Skinner, C. Wang, C. Colyer, J. Harrison, *Anal. Chem.*, **71**, 3036 (1999).
16. I.M. Lazar, R.S. Ramsey, S. Sundberg, J.M. Ramsey, *Anal. Chem.*, **71**, 3627 (1999).
17. N.H. Bing, C. Wang, C.D. Skinner, C. Colyer, P. Thibault, J.D. Harrison, *Anal. Chem.*, **71**, 3292 (1999).
18. B. Zhang, H. Liu, B.L. Karger, F. Foret, *Anal. Chem.*, **71**, 3258 (1999).
19. F. Xiang, Y.H. Lin, J. Wen, D.W. Matson, R.D. Smith, *Anal. Chem.*, **71**, 1485 (1999).

20. L. Licklider, X. Wang, A. Desai, Y. Tai, T.D. Lee, *Anal. Chem.*, **72**, 367 (2000).
21. J. Kameoka, R. Orth, B. Ilic, D. Czaplewski, T. Wachs, H.G. Craighead, *Anal. Chem.*, **74**, 5897 (2002).
22. M. Brivio, R.H. Fokkens, W. Verboom, D.N. Reinhoudt, N.R. Tas, M. Goedbloed, A. van den, *Anal. Chem.*, **74**, 3972 (2002).
23. S. Le Gac, S. Arscott, C. Rolando, *Electrophoresis*, **24**, 3640 (2003).
24. K. Huikko, P. Ostman, K. Grigoras, S. Tuomikoski, V.M. Tiainen, A. Soininen, K. Puolanne, A. Manz, S. Franssila, R. Kostiainen, T. Kotiaho, *Lab. Chip*, **3**, 67 (2003).
25. I.M. Lazar, L. Li, Y. Yang, B.L. Karger, *Electrophoresis*, **24**, 3655 (2003).
26. G.A. Schultz, T.N. Corso, S.J. Prosser, S. Zhang, *Anal. Chem.*, **72**, 4058 (2000).
27. S. Zhang, C.K. Van Pelt, J.D. Henion, *Electrophoresis*, **24**, 3620 (2003).
28. T.J. Kauppila, P. Ostman, S. Marttila, R.A. Ketola, T. Kotiaho, S. Franssila, R. Kostiainen, *Anal. Chem.*, **76**, 6797 (2004).
29. S. Arscott, S. Le Gac, C. Druon, P. Tabourier, C. Rolando, *Sensors & Actuators B*, **98**, 140 (2004).
30. J. Carlier, S. Arscott, V. Thomy, J.C. Camart, C. Cren-Olive, S.J. Le Gac, *J. Chromatogr. A*, **1071**, 213 (2005).
31. M. Schilling, W. Nigge, A. Rudzinski, A. Neyer, R. Hergenröder, *Lab. Chip*, **4**, 220 (2004).
32. M.F. Bedair, R.D. Oleschuk, *Anal. Chem.*, **78**, 1130 (2006).
33. S. Ssenyange, J. Taylor, D.J. Harrison, M.T. McDermott, *Anal. Chem.*, **76**, 2393 (2004).
34. S. Thorslund, P. Lindberg, P.E. Andren, F. Nikolajeff, J. Bergquist, *Electrophoresis*, **26**, 4674 (2005).
35. Y.N. Yang, J. Kameoka, K.H. Lee, H.G. Craighead, *Lab. Chip*, **5**, 869 (2005).
36. G.E. Yue, M.G. Roper, E.D. Jeffery, C.J. Easley, C. Balchunas, J.P. Landers, J.P. Ferrance, *Lab. Chip*, **5**, 619 (2005).
37. M. Brivio, R.E. Oosterbroek, W. Verboom, A. van den Berg, D.N. Reinhoudt, *Lab. Chip*, **5**, 1111 (2005).
38. J. Xie, Y. Miao, J. Shih, Y.C. Tai, T.D. Lee, *Anal. Chem.*, **77**, 6947 (2005).
39. A.V. Willems, D.L. Deforce, C.H. Van Peteghem, J.F. Van Bocxlaer, *Electrophoresis*, **26**, 1412 (2005).
40. M. Moini, H. Huang, *Electrophoresis*, **25**, 1981 (2004).
41. W.F. Smyth, *Electrophoresis*, **27**, 2051 (2006).
42. S. Amon, A. Plematl, A. Rizzi, *Electrophoresis*, **27**, 1209 (2006).
43. W. Kolch, C. Neususs, M. Pelzing, H. Mischak, *Mass Spectrom. Rev.*, **24**, 959 (2005).
44. B. Zhang, F. Foret, B.L. Karger, *Anal. Chem.*, **73**, 2675 (2001).
45. T. Razunguzwa, *Methods Mol. Biol.*, **339**, 67 (2006).

46. S. Akashi, K. Suzuki, A. Arai, N. Yamada, E.I. Suzuki, K. Hirayama, S. Nakamura, Y. Nishimura, *Rapid Commun. Mass Spectrom.*, **20**, 1932 (2006).

47. Y. Tachibana, K. Otsuka, S. Terabe, A. Arai, K. Suzuki, S. Nakamura, *J. Chromatogr. A*, **1025**, 287 (2004).

48. J.S. Kim, D.R. Knapp, *Electrophoresis*, **22**, 3993 (2001).

49. B. Wang, Z. Abdulali-Kanji, E. Dodwell, J.H. Horton, R.D. Oleschuk, *Electrophoresis*, **24**, 1442 (2003).

50. M. Svedberg, M. Veszelei, J. Axelsson, M. Vangbo, F. Nikolajeff, *Lab. Chip*, **4**, 322 (2004).

51. A.P. Dahlin, M. Wetterhall, G. Liljegren, S.K. Bergstrom, P. Andren, L. Nyholm, K.E. Markides, J. Bergquist, *Analyst*, **130**, 193 (2005).

52. C.E. Hop, Y. Chen, L.J. Yu, *Rapid Commun. Mass Spectrom.*, **19**, 3139 (2005).

53. T. Toyohide, S. Masuoka, J.Y. Jin, *J. Separation Sci.*, **26**, 635 (2003).

54. L.M. Holland, J.W. Jorgenson, *Anal. Chem.*, **67**, 3275 (1995).

55. K. Tsukagoshi, M. Hashimoto, T. Suzuki, R. Nakajima, A. Arai, *Anal. Sci.*, **17**, 1129 (2001).

56. M. Hashimoto, K. Tsukagoshi, R. Nakajima, K. Kondo, A. Arai, *J. Chromatogr. A*, **867**, 271 (2000).

57. A. Arora, J.C.T. Eijkel, W.E. Morf, A. Manz, *Anal. Chem.*, **73**, 3282 (2001).

58. K. Shin, Y.H. Kim, J.A. Min, S.M. Kwak, S.K. Kim, E.G. Yang, J.H. Park, B.K. Ju, T.S. Kim, J.Y. Kang, *Anal. Chim. Acta*, **573**, 164 (2006).

59. P. Schulze, M. Schüttpelz, M. Sauer, D. Belder, *Lab. Chip*, **7**, 1841 (2007).

60. A.M. Jorgensen, K.B. Mogensen, J.P. Kutter, O. Geschke, *Sensors & Actuators B*, **90**, 15 (2003).

61. D.J. Jackson, J.F. Naber, T.J. Roussel, J.M.M. Crain, K.M. Walsh, R.S. Keynton, R.P. Baldwin, *Anal. Chem.*, **75**, 3643 (2003).

62. J.S. Mellors, J.W. Jorgenson, *Anal. Chem.*, **69**, 983 (1997).

63. J.E. MacNair, K.D. Patel, J.W. Jorgenson, *Anal. Chem.*, **71**, 700 (1999).

64. M. McEnery, A. Tan, J. Alderman, J.D. Glennon, J. Alderman, J. Patterson, S.C. O'Mathuna, *Analyst*, **125**, 25 (2000).

65. R. Wilke, S. Buttgenbach, *Biosen. Bioelectron.*, **19**, 149 (2003).

66. K.T. Liao, C.M. Chen, H.J. Huang, C.H. Lin, *J. Chromatogr. A*, **1165**, 213 (2007).

67. Y.T. Hsueh, R.L. Smith, M.A. Northrup, *Sensors & Actuators B*, **33**, 110 (1996).

68. A. Arora, A.J. de Mello, A. Manz, *Anal. Commun.*, **34**, 393 (1997).

69. R.M. Guijt, E. Baltussen, G. van der Steen, H. Frank, H. Billiet, T. Schalkhammer, F. Laugere, M. Vellekoop, A. Berthold, L. Sarro, G.W. van Dedem, *Electrophoresis*, **22**, 2537 (2001).

70. F. Laugere, R.M. Guijt, J. Bastemeijer, G. van der Steen, A. Berthold, E. Baltussen, P. Sarro, G.W. van Dedem, M. Vellekoop, A. Bosschet, *Anal. Chem.* **75**, 306 (2003).

71. L. Jiang, Y. Lu, Z. Dai, M. Xie, B. Lin, *Lab. Chip*, **5**, 930 (2005).

72. Q.J. Song, G.M. Greenway, T. McCreedy, *J. Anal. Atom. Spectrom.*, **18**, 1 (2003).

73. Q.J. Song, G.M. Greenway, T. McCreedy, *J. Anal. Atom. Spectrom.*, **19**, 883 (2004).

74. F. Li, D.D. Wang, X.P. Yan, J.M. Su, R.G. Lin, *Electrophoresis*, **26**, 2261 (2005).

75. A.Y.N. Hui, G. Wang, B. Lin, W.T. Chan, *J. Anal. Atom. Spectrom.*, **21**, 134 (2006).

76. H.Y. Aboul-Enein, I. Ali, *Chiral Separations by Liquid Chromatography and Related Technologies*, New York: Marcel Dekker (2003).

77. I. Ali, H.Y. Aboul-Enein, *Chiral pollutants: Distribution, toxicity and analysis by chromatography and capillary electrophoresis*, Chichester: Wiley (2004).

78. I. Ali, H.Y. Aboul-Enein, *Instrumental methods in metal ions speciation: Chromatography, capillary electrophoresis and electrochemistry*, New York: Taylor & Francis (2006).

79. J.P. Chervet, M. Ursem, J.P. Salzmann, *Anal. Chem.*, **68**, 1507 (1996).

80. S. Mayer, X. Briand, E. Francotte, *J. Chromatogr. A*, **875**, 331 (2000).

81. K. Uchiyama, A. Hibara, K. Sato, H. Hisamoto, M. Tokeshi, T. Kitamori, *Electrophoresis*, **24**, 179 (2003).

82. F. Kitagawa, T. Tsuneka, Y. Akimoto, K. Sueyoshi, K. Uchiyama, A. Hattori, K. Otsuka, *J. Chromatogr. A*, **1106**, 36 (2006).

83. F. Laugere, G.W. Lubking, J. Bastemeijer, M. Vellekoop, *Sensors & Actuators B*, **83**, 104 (2002).

84. K.B. Mogensen, N.J. Petersen, J. Hübner, J.R. Kutter, *Electrophoresis*, **22**, 3930 (2001).

85. K.W. Ro, K. Lim, B.C. Shim, J.H. Hahn, *Anal. Chem.*, **77**, 5160 (2005).

86. B. Ma, X. Zhou, G. Wang, Z. Dai, J. Qin, B. Lin, *Electrophoresis*, **28**, 2474 (2007).

87. X. Li, Z. Wen, Z. Wen, *Sepu*, **22**, 469 (2004).

88. A.D. McBrady, R.E. Synovec, *J. Chromatogr. A*, **1105**, 2 (2006).

CHAPTER 5

SAMPLE PREPARATION IN NANOCHROMATOGRAPHY AND NANOCAPILLARY ELECTROPHORESIS

5.1. INTRODUCTION

Generally, drugs and xenobiotics are present in very low concentrations in biological and environmental samples and, therefore, their analysis is a tedious job. Besides, thousands of compounds, as impurities, are present along with species of interest in unknown matrices. Samples containing high ionic matrices cause problems in nanocapillary electrophoresis (NCE) as high ionic strength imparts a low electric resistance resulting in very poor, broad peaks. In addition to this, electroosmotic flow (EOF) in the capillary is altered by the influence of the sample matrix, which may result in poor separation. Sometimes, the detector baseline is perturbed when the pH of the sample differs greatly from the pH of the background electrolyte (BGE). Samples containing UV-absorbing impurities may be problematic in detection. Sometimes, detection limits of certain species have been increased by using preconcentration methods. Besides, enantiomers of drugs, xenobiotics, and metal ion species (of the same metal ion) may have similar properties, which again make analysis difficult and require special sample preparation methods. Moreover, detection is the major drawback in nanoliquid chromatography (NLC) and NCE because of the low sample quantity or samples having

Nanochromatography and Nanocapillary Electrophoresis. By Ali, Aboul-Enein, and Gupta
Copyright © 2009 John Wiley & Sons, Inc.

poor ingredient concentrations. Under such circumstances, the sample preparation is essential and required before starting analysis in NLC and NCE.

One of the most important trends to simplify these complications is the generation of simple, rapid, and reliable procedures for sample preparation. Method development and setup require the use of materials of known compositions, for example, certified reference materials. Therefore, spiking experiments have to be performed for method quality control. Under such experimental conditions, emphasis has to be placed on the spiking procedures as they exert an influence on the recovery values. Although present scientific knowledge is not perfect, the use of spiking experiments helps to minimize the errors. The integration and automation of all steps between sample preparation

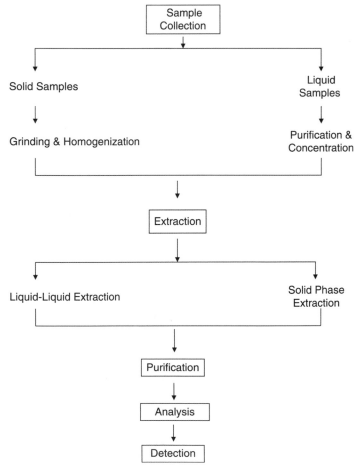

Figure 5.1   General protocol for sample preparation.

and detection significantly reduce the time of analysis, increasing reproducibility and accuracy. Therefore, extraction, purification, and preconcentration of real samples are very important in NLC and NCE. The latest developments in sample handling have been highlighted in several conferences and some reviews have been presented on this issue [1–19]. A general protocol of sample preparation is given in Fig. 5.1. This chapter describes state-of-the-art sample preparation, including sampling, extraction, purification, and preconcentration processes required before NLC and NCE analyses.

5.2. SAMPLE PREPARATION

We have carried out an extensive literature search on sample preparation technologies and found many papers on conventional chromatography and capillary electrophoresis methods but few on NLC and NCE. It is important to mention here that sample preparation methodologies used in conventional chromatography and capillary electrophoresis can be used in NLC and NCE. The interested reader can consult our earlier books for details [20,21]. However, attempts have been made to describe sample preparation protocols required in NLC and NCE techniques. Some of the important requirements and preparations are discussed below.

5.3. SAMPLING

Before starting work it is necessary to design the sampling protocol. It is the first and most important step in NLC and NCE and its design and implementation has a decisive influence on the analyses. Due to low sample volume and poor ingredient concentration careful collection of the sample is required. Before sampling, complete knowledge and information about surroundings and physiology is needed. Changes in physicochemical properties (transfer, transformation, deposition, accumulation, etc.) of the target species have to be considered in more detail. By considering these points useful samples can be composed, which may significantly improve the analytical results. Some important facts and issues needed in biological and environmental samples are discussed below.

5.3.1. Biological Samples

For biological sample collection, the history of the animal, treatment given, if any, and dose, and the surroundings in the case of plants should be known in detail. Biological rhythmicity is very important to consider at the time of biological sampling. During sampling some important precautions should be adopted, which include avoidance of cloth, wool, contacts, etc. There should be no use of acids or bases at the sampling site and all the equipment should be neat and clean. The biosphere varies greatly in its physical and chemical properties relevant to species distribution and transformation compared to abiotic environments such as air, water, sediment, and soil. A detailed sampling protocol has to be developed and adopted separately for the specific studies depending on the particular requirement. Sampling strategies and procedures for some biological samples are summarized below.

5.3.1.1. Blood   The representativeness with reliable methodology is very important for blood samples. The skin area for blood collection should be cleaned with deionized water followed by ethanol or methanol and should be allowed to evaporate properly. Blood samples are collected by ETS systems; however, plastic syringes have also been used for collection of blood samples. Microdialysis is a newly developed technique and has been used for the collection of blood samples. Due to the risk of contamination, a syringe made of ordinary glass should be avoided. The addition of anticoagulant into the collected blood sample should be carried out very carefully as these have good capacity to complex with metal ions and other molecules, and, therefore, a blank for each species should be run. Generally, blood samples are stored at $-20°C$ (for 24 h).

5.3.1.2. Urine   Much attention is required in urine sampling due to the greater possibility of contamination and, hence, more skill is required than for a blood sample. The urine sample can be collected directly or from the urinary bladder. The samples can be collected with the help of an ETS system or syringe or a measuring cylinder made of plastic. The collected samples can be kept in polyethylene containers that can be closed with airtight lids. The samples should be analyzed immediately or preserved at low temperature, if analysis will take some time.

5.3.1.3. Tissues   Tissue sampling depends on the type of analysis and, normally, it is carried out from liver, kidney, the gastrointestinal tract, and other body parts. The probability of contamination in tissue sampling is quite high due to the long processes involved. The sampling of tissues is carried out by microdialysis, needle aspiration, or biopsies (with surgical cutting apparatus). Generally, tissues are washed with 1.0% salt solution and kept in plastic containers at low temperature. Again analysis should be completed immediately or samples should be preserved at low temperature if analysis will be delayed.

5.3.1.4. Plants
Sampling of plants can be carried out simply by cutting the part of interest. The sampling is carried out by means of a knife, scissors, etc., in higher plants. A sample from terrestrial plants can be obtained after washing the outer cell wall of the plant with deionized water. Required samples are collected by dissecting the specific plant parts. The specific samples such as from xylem, phloem, etc., are collected carefully by the dissection of the plant stem. Samples from root are collected by digging, washing, and dissecting the roots properly. Similarly, samples from aquatic plants can be collected with the help of scissors, a knife, blade, forceps, etc. In the case of phytoplanktons sampling can be done using a net.

5.3.1.5. Food Stuffs
The choice of sampling for food stuffs depends on the type of analysis and sampling methods. Sample collection from food stuffs should be carried out carefully as there are greater chances of contamination. Samples of beverages are collected with the help of a syringe or cylinder made of inert plastic. Samples from solid food stuffs are collected with the help of a spoon, knife, blade, forceps, etc. The collected samples can be preserved by adding some acids and can be stored at low temperature for a few months. For more details readers should consult the sampling protocol for food stuffs provided by the *Official Control of Foodstuffs Directives* [22–24].

5.3.2. Environmental Samples

Environmental sampling protocol and strategy include selection of sampling sites, type of sample (grab, mixed, or composite), sample container, volume collection, sample handling, transportation, preservation, and storage. The selection of the sampling sites and the type of sample are based on the objectives and the nature of the study. The sampling strategies for air, water, soil, and sediment samples are summarized below.

5.3.2.1. Air
The sampling requirement depends on type of analysis; normally, air samples are collected for the analysis of pollutants. Generally, air is sampled for dust, smoke, and organic and inorganic pollutants. Air samples are collected into balloons, cylinders, canisters, glass traps, etc. Vacuum filling of stainless steel containers is the official sampling method prescribed by the U.S. Environmental Protection Agency for volatile compounds in monitoring urban air [25]. In addition, cryotrapping, solid adsorbent cartridges, polymer bags, and stainless steel canisters are used for air samplings. Aerosols are

collected using impactors, filters, denuders, electrostatic separators, etc. [26]. Air particles are also sampled using filters, glass fiber, the Hi-Vol sampler, the Institute of Occupational Medicine sampler, seven holes sampler, etc. Personal exposure monitors (PEMs) are sampling devices worn on the body to estimate an individual's exposure to air pollution. Porous polymers, such as Tebax-GC, XAD resins, and polyurethane forms, have been used in gas sampling. Other sorbents used for this purpose are charcoal, carbon molecular sieves, etc. It is very important to mention here that the inner wall should be inert [coated with polymers such as polytetrafluoroethylene (PTFE)]. The gas phases can be collected into a pre-evacuated sampling device [27]. Air is passed through a 0.45 μm filter to remove solid particles and aerosols up to a predefined size from the sample. Generally, sample collection involves an air inlet, a particulate separator or collecting device, an air flow meter, and a flow rate control valve air pump. Important parameters to be considered during sampling include temperature, light intensity, humidity, oxygen content, and aerosol concentration. Boiano et al. [28] compared the sampling efficiency of three methods, the OSHA Method ID-215, NIOSH Method 7605, and NIOSH Method 7703. Roinestad et al. [29] improved the sampling of indoor air and dust using a Tenax TA sampling pump.

5.3.2.2. Water   Sampling methods for water vary depending on the nature of the compounds and the type of water, that is, ground, waste, lake, river, sea, tap, industrial, soil, rain, and snow water. The location and depth of the water sample collection site also influence sampling methods. Water samples having volatile species are collected by an online combination of purging water with an inert gas such as helium followed by cryogenic trapping. Polycarbonate or polyethylene bottles are recommended for most water sample collection. To avoid any adsorption on glass bottles these should be pretreated with nitric acid. The containers should be washed properly with tap water, acids (no plastic bottles), and detergents, then again with tap water and double distilled water.

Sample collection varies from hand sampling procedures at a single point to more sophisticated multipoint sampling techniques, such as the equal discharge increment (EDI) method or equal transit rate (ETR) method. Different types of water samplers can be used to collect water samples from bodies such as rivers, seas, lakes, etc. The Blumer sampler [30], Deutsche Hydrographisches Institute sampler (DHI) [31], and a high volume water sampler, designed to pump water from a defined depth below the water body surface outside the wake of the survey vessel, are suitable as samplers [32]. The Kemmerer and van Dorn samplers may also be used for water sample collection. For rivers, up- and downstream samples are usually required, as well as the points where tributaries or waste drains join the main stream under study.

The rigorous clean-up of all parts of the sampler and apparatus, solvent, adsorbent materials, and other chemicals should be carried out very carefully. Possible contamination due to penetration of the sampler through the surface layers of the water body, which may be highly enriched by the pollutants, has to be excluded. The environmental samples collected should be transported to the laboratory immediately and analysis should start as soon as possible.

5.3.2.3. Sediment and Soils Sediment is the most important part of a riverine system by which contamination levels can be measured exactly. Representativeness is very important in sampling of sediment and soil. Soil and sediment samples may be collected with the help of samplers such as the Ponar grab, Orange peel, Smith McIntyre grab, Shipek grab, Petersen grab, vanVeen grab, Ekman grab, etc. [33]. The texture of soil and sediment is heterogeneous and an extensive screening of each sampling site is necessary for accurate scientific analyses. Numbers and distances of the lateral and vertical sampling points are selected on the basis of geological information by applying statistical models. Related sampling grids and procedures are available in the literature [34–36]. Sampling of sediment and soil is carried out using a shovel, corer, spoon, or knife made of titanium, ceramics, or plastics. Soil can be collected by initial separation of mineral layers or as intact depth profiles. The latter approach is very common in the sampling of lake or marine sediments, which is achieved using different types of corer. Sediment can also be collected by grab or core samplers. Sediment traps are used in dynamic flow systems, such as a river, culvert, or canal. Special procedures for the species retaining sampling of soil and sediment are still lacking in the literature. Problems also occur from change in the oxygen concentration during sampling, especially for originally anoxic sediments. The interaction between solid particles and water, biofilms and pore water occur and, hence, definition of the sample composition is difficult for a variety of purposes. Therefore, the development of *in situ* methods and new approaches to species retaining sampling are required.

5.4. PRESERVATION

Sometimes, the laboratories and instruments are not available immediately after sampling, especially when the sampling is carried out at remote areas. Under such conditions, storage of the collected samples is required. Preservation of the chemical identities of the sample is the prime objective. Selection of the preservation method depends on the type of sample, storage time, size, and number of samples. The temperature conditions for preserving biological samples vary from 0°C to −130°C or more depending on the type of sample collected. Samples can be stored for several years [37] by keeping them in special cryocontainers in gas phase above liquid nitrogen. Sometimes, water is removed from biological tissues to prevent bacterial growth but this method of storage is not universal and effective as some species may be removed with water. Sometimes, in the case of urine, precipitation occurs and, hence, these types of samples should be stored in the presence of preservatives (0.03 M/L HCl or HNO_3, sulfamic acid triton-X-100). Care should be taken to adjust the pH of urine as some metal-protein complexes are destabilized at lower pH. Briefly, the storage of the environmental and biological samples at low temperature (0°C to −130°C) is the best method of sample preservation.

Air samples cannot be stored directly for future analyses; however, the preservation of gaseous components in a cold trap is possible for one week or more in liquid nitrogen. The storage of gaseous samples for a long time is not recommended. Certain species get lost due to adsorption on the container or ion exchange with the glass container. To avoid adsorption and exchange, concentrated nitric acid (1.5 mL/L or pH <2) should to added to water samples prior to storage. But 5.0 mL/L concentrated nitric acid should be used in samples of high buffer capacity. These water samples should be stored in an incubator at about 4°C to avoid evaporation. Samples containing microgram per liter concentrations should be analyzed immediately. Sometimes, certain special preservatives may be used to store the samples for inorganic and organic pollutants. The acidification of water or other liquid sample is carried out to stop adsorption and exchange processes but it is not suitable if the sample contains organometals, which may become depredated in acidic conditions. Sediment and soil samples are stored normally at 0°C.

5.5. FILTRATION

Generally, liquid biological and environmental samples cannot be analyzed directly and need some processing. Urine, fluids, and water samples may contain some solid impurities and they require filtration to extract them before proceeding. Filtration of these samples can be carried out using a vacuum pump and Buchner funnel assembly with Whatman's filter paper. Weigel et al. [38] described a PTFE tubing fitted with a gear pump (model MCP-Z pump head Z-120 with PTFE gears, magnet 66, Ismatec, Wertheim, Germany) for the filtration of liquid samples. The pump can be placed behind the extraction unit for calculating the recovery rates from purified water. For online filtration of the sample glass fiber filter candles (height 82 mm, outer diameter 26 mm,

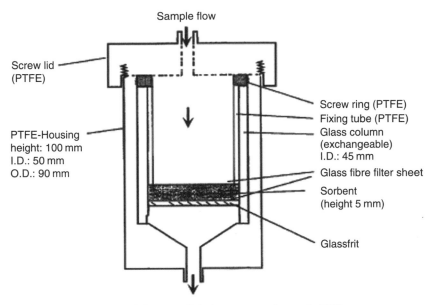

Figure 5.2 A solid phase extraction unit [38].

inner diameter 14 mm) were used in a stainless steel housing, as shown in Fig. 5.2 [38]. Filtration is not required for solid samples; solid samples can be manipulated if they contain other solid materials. If dust particles are present in an air sample they should be removed before loading the collected air into the machine. Filtration of the air sample is carried out using perch membranes under pressure.

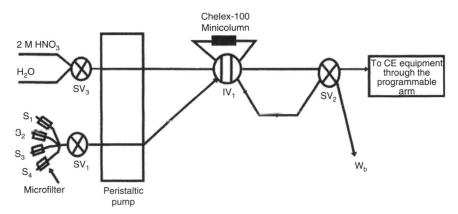

Figure 5.3 Single coupling flow injection system for sample preparation introduction in CE system [39].

Arce et al. [39] developed a flow injection analysis (FIA) system (Fig. 5.3) for online filtration of water samples prior to CE analysis. They also constructed a pump-driven unit for extraction and filtration of soil samples combined with CE in an online mode (automated sample transfer between pre-CE sample preparation step and the CE) [40]. The method was precise and four times faster than conventional methods of sample preparation with an off-line unit. Blood samples are centrifuged immediately to remove red blood cells and the serum is stored as discussed above. Sometimes, urine samples also contain precipitates which are removed by centrifuge.

5.6. DIGESTION/HOMOGENIZATION

Normally, solid biological and environmental samples need digestion and homogenization before analysis. The treatment of solid samples is more complicated in comparison to liquid or gas samples. In biological samples, some molecules remain bonded with proteins and other biomolecules, which require a rigorous extraction procedure. Since 1970, acid homogenization methods have been used for the extraction of compounds from solid biological and environmental samples [4]. Toluene, methanol, etc., are suitable solvents for homogenization of biological samples but, sometimes, dilute acids along with the above-mentioned solvents may be used, under special conditions. For organometalics, the environmental and biological samples are digested with acid or base which are extracted by organic solvents. Normally, concentrated nitric acid is used for homogenization of soil and sediment samples but, sometimes, mixtures of nitric acid, sulfuric acid, hydrochloric acid, perchloric acid, etc., are used to digest the samples.

Generally, digestion is carried out in a flask or beaker by heating on a hot plate. Nowadays, a homogenizer is used for this purpose. The homogenizer contains small sets of blades and generator probes that cause vigorous mixing and turbulence, which optimizes sample contact and preparation. The homogenization of biological and environmental samples is tedious and time consuming. Homogenization is aided by shaking these samples at different speeds. Sometimes, mild heating also improves extraction of molecules during the homogenization procedure. The direct or indirect sonication of the samples helps greatly to extract species of interest from solid samples. Sonication causes a disruption of the cell wall in biological samples, that is, plant cells, bacteria, fungi, and yeast and it helps in the release of the analytes from these biological solid samples. After disruption, the proteins, DNA, RNA, and other cellular components can be separated for additional purification or processing.

5.7. EXTRACTIONS

The extraction is the most important step in sample preparation of biological and environmental matrices. Many devices have been used for the extraction procedures. The most important techniques used before conventional chromatography and capillary electrophoresis analyses include:

Extraction of solid samples
Homogenization extractions
Sonication extractions
Microwave assisted extractions
Soxhlet extractions
Accelerated solvent extractions
Supercritical fluid extractions
Extraction of liquid samples
Liquid-liquid extractions
Solid phase extractions (SPE)
Liquid chromatographic extractions
Column chromatography
High performance liquid chromatography (HPLC)
Supercritical fluid chromatography (SFC)
Gel permeation chromatography (GPC)
Membrane methods in sample preparation
Enzymatic hydrolysis

These methods are fully developed for conventional chromatography and capillary electrophoresis. They are not discussed in detail here and the interested reader should consult our earlier books [20,21]. However, few papers are available on those techniques capable of working at nanolevel analysis and these have been summarized in this chapter.

5.8. CLEAN UP

In the case of off-line sample preparation methods, clean up is required in samples having colors and other interfering impurities. Biological and environmental samples contain thousands of other substances as impurities and these were extracted along with the analytes of interest. Due to the similar properties of the co-extractives, they usually interfere in the analyses of the interested molecules. Therefore, a clean-up procedure is required only for those samples that are not clear. Blood, serum, food, plant extract, and industrial and municipal effluents require a clean-up procedure. If clean up is required, the recovery of each constituent is very important and it should be at about 85% [41]. Clean up of the extracted species may be carried out by column chromatography, gel permeation chromatography, sweep co-distillation, liquid-liquid partition, cartridges and disks. The columns used for this purpose are nuchar carbon, silica, XAD-2, alumina, flurosil, while the cartridges and disks used are C_8, C_{18}, etc. The use of membrane disks for clean up has increased significantly in the past few years. SPE membrane disks may also be used in conjunction with supercritical fluid extraction in the preconcentration of different molecules. Online SPE may be coupled to membrane disk for the clean up of species in biological and water samples. Good recovery of the extracted species can be achieved using a SFC/SPE methodology. Some clean-up methods for various species are given in Table 5.1.

TABLE 5.1   Some Clean-Up Methods for Different Molecules

Method	Co-Extractive	Nature
1. Partitioning		
a: Liquid-liquid partitioning	Fats, waxes, lipids, pigments, & some polar compounds	General
b: Partitioning adsorption column	Fats, phospholipids, waxes and other polar substances	General
2. Gel permeation chromatography	Fats & lipids	General

3. Liquid-liquid adsorption chromatography (adsorption column)
   a: Flurosil
   b: Alumina
   c: Silica gel
   d: Nuchar carbon
	Fats, phospholipids, pigments, polar impurities	General

4. Chemical clean up
 	
   a: Acidic
   b: Alkaline
   c: Acid-base
	Fats & neutral specific molecules	—

5. Sweep co-distillation	Comparatively nonvolatile impurities	—

5.9. PRECONCENTRATION

NLC and NCE deal with low amounts of sample, which requires preconcentration procedures, especially in the case of off-line sample preparation methods. The volume of the extracted solvent containing species is reduced to 1 to 10 μL. The classical approach of evaporation of the solvents is used to concentrate the extracted species. Some devices, such as the purge and tap device and cryogenically cooled capillary taps, have been used for preconcentration [42]. Solvents with low boiling points are the best suited for preconcentration. Open evaporation of extracts on a water bath may cause severe loss of the molecules, and, therefore, evaporation should be carried out at low temperature under reduced pressure. Normally, a rotary flash evaporator or Kuderna-Danish assembly is used for this purpose. Sometimes, pure nitrogen is also allowed to pass through the evaporator assembly so that oxidation of the species may be avoided. Sometimes, loss of the molecules occurs during evaporation and to check this some keeper solvents (toluene or iso-octane, which have high boiling points) are added to the extract before starting the evaporation. Besides this conventional method, other techniques such as chromatography and SPE have also been used for preconcentration. In addition to this, amalgam formation, cold trap, high temperature trap, etc., have been used for preconcentration.

During extraction, clean up, and preconcentration (especially in LLE methodology) some invisible moisture may remain in the sample along with organic solvents which may create problems in detection of the analytes. Therefore, moisture should be removed from the sample prior to its loading onto the analytical machine, especially on GC. Normally, moisture from organic solvent is removed by addition of anhydrous sodium sulfate of high purity. Sometimes, the extracted and concentrated solvent is passed through a glass column containing anhydrous sodium sulfate.

5.10. OFF-LINE NANOSAMPLE PREPARATION METHODS

The various sample preparation methods mentioned in Section 5.6 have been tried in microfluidic devices off-line for biological and environmental matrices followed by analysis of the extracted samples by NLC and NCE. The available papers in the literature on this subject are discussed in the following sections.

5.10.1. Nano Solid Phase Extractions

Nowadays, solid phase extraction (SPE) is the technique of choice because it is inexpensive, fast, reproducible, and has a wide range of applications [43,44].

Therefore, more than 50% of the analytical scientists are using this method of sample preparation. During extraction of biological and environmental samples, both biogenic and anthropogenic compounds are extracted into organic and inorganic solvents. The complex matrix may interfere seriously with the determination of the respective analytes. Various columns, disks, and cartridges have been used for extraction but cartridges are the most popular extraction devices [45]. The manufacturers have developed new formats for traditional cartridges. Most SPE cartridges have medical grade polypropylene syringe barrels with porous PTFE or metal frit that contain 40 μm d_p packings. The cartridges provide certain advantages over disks as liquid flows faster in cartridges in comparison to disks. Most cartridges and disks are silica-based polymeric packing. Nano-SPE microfluidic devices contain adsorbent inside, that is, reversed phase silica and other polymers. Few papers have been found in the literature on nano-SPE; they are discussed here.

Jemere et al. [46] described a 2000-fold concentration of peptides using nano-SPE on octadecylsilane-packed columns, which allowed sub-picomolar detection in NCE. Ekström et al. [47] developed a silicon microextraction chip (SMEC) for sample clean up and trace enrichment of peptides. The authors validated this device successfully carrying out analysis on a 10 nM peptide mixture containing 2 M urea in 0.1 M phosphate buffer. Furthermore, the authors used this to trap beads immobilized with trypsin, effectively making it a microreactor for enzymatic digestion of proteins. This microreactor was used to generate a peptide map from a 100 nM bovine serum albumin sample. Jandik et al. [48] described a microfluidic device using a laminar fluid diffusion interface for blood sample preparation. The authors reported this method was useful in HPLC by eliminating the need for centrifugation and reducing sample preparation time from 30 to 60 min to 5 min. Kuban et al. [49] presented a miniaturized continuous ion exchanger for sample preparation. Xu et al. [50] used a photoactivated polycarbonate solid phase reversible immobilization device for purification of DNA, especially for the removal of dye terminator and other soluble components in medium. Breadmore et al. [51] reported the purification of DNA from blood in less than 15 min on a microchip with immobilized silica beads. Broyles et al. [52] described a microdevice having C_{18} as a stationary phase and on-chip filtration by an array of thin channels to exclude particulates and sample concentration. Polynuclear aromatic compounds were analyzed by electrochromatography. Gustafsson et al. [53] described high throughput microfluidic concentration of protein digests integrated with matrix-assisted laser desorption/ionization (MALDI) mass spectrometry on a chip. According to the authors, centrifugal forces move liquid through multiple microstructures, each containing a 10 nL C_{18} column. The samples are concentrated, desalted, and subsequently eluted from the columns directly into MALDI target areas ($200 \times 400\ \mu m$).

Hofmann et al. [54] described a method for preconcentration of spores on a polydimethylsiloxane (PDMS) target plate. Bozalongo and coworkers [55] optimized solid phase microextraction for analysis of volatile compounds in ground wood samples. The authors studied this method in terms of repeatability and extraction efficiency in samples of American and French oak with different degrees of toasting. Chiesl et al. [56] purified protein from DNA using water-soluble block copolymers of acrylamide and N-alkylacrylamides. Bhattacharyya et al. [57] reported a polymeric SPE for isolation of DNA using plastic chip as a disposable sample preparation system. The extraction was achieved due to binding of nucleic acids to the silica particles. The solid phase was formed within the channels of the device with *in situ* photoinitiated polymerization of a mixture of methacrylate and dimethacrylate monomers. Huynh et al. [58] integrated sample derivatization steps on a chip using naphthalene 2,3-dicarboxaldehyde and 2-mercaptoethanol involving microdialysis and NCE. The chip was made of a glass layer etched with microfluidic channels and sealed with a layer of PDMS via plasma oxidation. The unit was used for analysis of a mixture of amino acids and peptide derivatives. Shiddiky and Shim [59] developed a sensitive on-chip preconcentration, separation, and detection of DNA in NCE. The chip contains three parallel channels, the first two for field-amplified sample stacking and subsequent field-amplified sample injection steps and the third for NCE. Buratti et al. [60] described solid phase extraction coupled with HPLC on polyurethane foam (PUF) chips for analysis of PAHs at nanogram per liter levels in urine samples. The authors described different parameters affecting analyte extraction on the PUF chip. The most important factors studied are urine salting-out, organic additives, clean up and desorption. Magnesium sulfate and tetrahydrofuran were added to 40 mL acidified urine sample and extracted by shaking at 30 rpm for 1 h at ambient temperature. The desorption was carried out after a clean up with diluted sodium hydroxide, using a small amount of diethyl ether. The authors reported percentage recoveries of $>90\%$ in the 2 to 100 ng/L range, with 0.1 to 0.5 ng/L as the detection limit.

5.10.2. Nano Membrane Extractions

The membrane technology has been tested in microfluidic devices. Normally a membrane is mounted between two chips, which make a microchannel, and fluid is allowed to pass through the membrane channel. Some papers are available on this method, which are discussed here. Hisamoto et al. [61] reviewed the application of capillary assembled microchips on PDMS as an online deproteinization device in NCE. These devices employed square section channels, for diffusion-based separation of small molecules from a sample having proteins. Lion et al. [62] presented a membrane desalting unit permitting the controlled elution of analytes from the membrane. The authors used this device for the detection of drugs, peptides, and proteins by electrospray ionization-MS. Similarly, Timmer et al. [63] described a sample preparation unit based on semipermeable membrane. Song et al. [64] presented a protein preconcentration unit using a 50 μm thick nanoporous polymer membrane on a chip. Foote et al. [65] described porous silica membrane-based (membrane between adjacent microchannels) sample preparation and preconcentration for proteins. Wang et al. [66] reported liquid membrane extraction and sample enrichment of haloacetic acids.

5.10.3. Nano Miscellaneous Extractions

Apart from SPE and membrane methods some other techniques were tested in microfluidic devices, which are also covered in this section. Nyholm [67] reviewed the application of chip-based electrochemical techniques for sample clean up, preconcentration, and derivatization. The authors also described electrochemical on-chip pumping, sample preparation, immuno-assays, and detection in NCE. Zhang and Timperman [68] reported preconcentration of fluorescein by exploiting an application of an electric field across the channel in NCE. Wong et al. [69] described an electrokinetic unit combined with electrophoretic and dielectrophoretic forces and the electrohydrodynamic flow for concentrating bioparticles from micrometer to nanometer in size. Wang et al. [70] described an electrokinetic trapping-based preconcentration device, which was capable of concentrating samples up to 108 times. Dhopeshwarkar et al. [71] reported an electrokinetic concentration of DNA using microchannels containing a hydrogel microplug, showing enrichment factors of 500 within 150 seconds. Leinweber et al. [72] discussed induced charge electrokinetics in structured electrode arrays. Li et al. [73] achieved a 10 to 100-fold enhancement of sample loading using electrokinetic injection of proteins/peptides from solution reservoirs.

Grass et al. [74] reported an isotachophoretic preconcentration prior to capillary electrophoretic separation and detected seleno amino acids. Xu et al. [75] used free-flow isoelectric focusing (IEF) for concentrating rhodamine 110 and fluorescein molecules at nanoliter sample volumes. Ross and Locascio [76] described a temperature gradient focusing for concentrating and separating ionic species in NCE. Authors reported an increase in concentrations up to 10,000 fold by balancing electrophoretic velocity of analyte against the bulk flow of solution in the presence of a temperature gradient. The method has

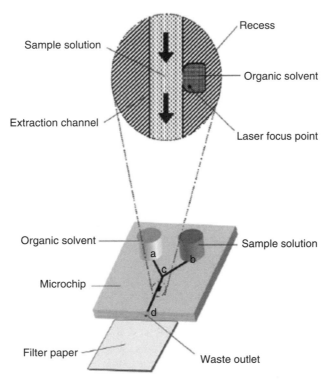

Figure 5.4 Schematic representation of liquid-liquid extraction on a chip with a square recess (150 μm long, 100 μm wide, and 25 μm deep with a shrunken opening of 50 μm width) fabricated in the channel walls of the chip [79].

been tested for fluorescent dyes, amino acids, DNA, proteins, and particles and was found capable of greater than 10,000-fold concentrations of the analytes. Holden et al. [77] reported a laminar microfluidic diffusion diluter to obtain concentration gradients. Liu et al. [78] reported sample stacking in isoconductive buffer systems due to ion transport mismatches, which caused changes in buffer conductivity in NCE. Chen et al. [79] developed a microchip liquid–liquid extraction for extraction and analysis of butyl rhodamine B. Organic solvent droplets were trapped in recesses fabricated in the channel walls, which get enriched with analyte. A schematic representation of this unit is shown in Fig. 5.4 indicating different channels on the chip.

5.11. ONLINE NANOSAMPLE PREPARATION METHODS

As discussed above, NLC and NCE are new development for the analysis of samples having low volume or poor analytes and, therefore, online sample preparation techniques are the best choice for accurate results. During the literature survey we found few papers dealing with online chip-based sample preparation methods in NLC and NCE. Attempts have been made to discuss these in the following paragraphs.

Huynh et al. [80] described the first hyphenation of a microdialysis NCE system for monitoring hydrolysis of fluorescein mono-β-D-galactopyranoside (FMG) by β-D-galactosidase. The layout of the microdialysis/microchip CE device is shown in Fig. 5.5 indicating channel lengths, voltage scheme, perfusate (20 mM sodium phosphate buffer, pH 7.4). Furthermore, the same authors [81] presented an online microdialysis sampling unit coupled with NCE. The authors used this set up for analysis of amino acid and peptide. A schematic representation is shown in Fig. 5.6, which shows the chip and microdialysis unit. Yazdi et al. [82] developed a novel method for microextraction of aromatic amines, coupled with NLC. The authors used hydrophobic porous polypropylene membranes as the interface between donor water sample and acceptor aqueous solution. The calibration curves for analytes were linear within the range of 20.0 ng/L to 300 µg/L. Long et al. [83] developed a microfluidic device integrating membrane-based sample preparation with NCE. The device consists of 10 nm pore diameter membranes sandwiched between two layers of PDMS substrates with embedded microchannels (Fig. 5.7). The microchannels were about 40 mm deep, 100 mm or 50 mm wide for upper and lower layers, respectively, and the size of the membrane was about 4 mm × 68 mm. Material exchange between two fluidic layers could be controlled precisely by applied voltages because of the membrane

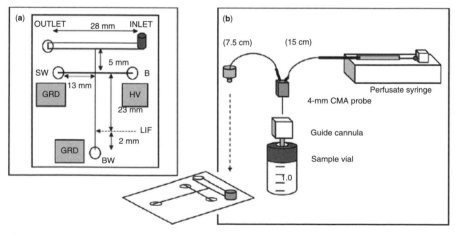

Figure 5.5   Schematic representation of online hyphenation of microdialysis sample preparation unit with NCE [80].

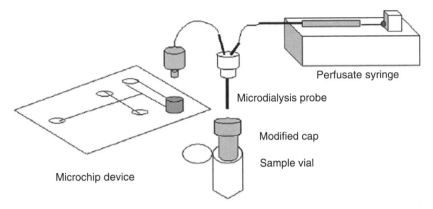

Figure 5.6   Schematic representation of microdialysis unit coupled with NCE [81].

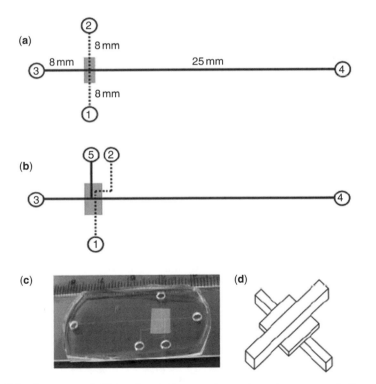

Figure 5.7   Layout and dimensions of a membrane-based preconcentration device. (a) Filter-CE unit, (b) concentrator-CE device, (c) dimension of microchannel, and (d) out look of concentrator-CE device [83].

isolation. Only small molecules passed through the nanopores and, hence, the integrated membrane served as a filter or a concentrator prior to NCE. The authors used this hyphenation for analysis of glutathione in human plasma and red blood cells without any off-chip deproteinization procedure. Furthermore, the same group [84] reported an integrated microfluidic device for online coupling of solid phase extraction to NCE. The preconcentration and separation were carried out simultaneously with a nanoporous membrane sandwiched between two PDMS substrates on a 2.5 mm long microcolumn, with two weirs on both sides to retain the C_{18}-coated silica beads. The authors claimed this unit might be considered a universal solution to online SPE-NCE hyphenation. This unit was tested for rhodamine 123 and FITC-labeled ephedrine and the authors claimed high separation efficiency and 1000-fold signal enhancement.

Wilson and Konermann [85] reported an online desalting of macromolecules (betaine-type amphoteric or zwitterionic surfactant solutions) using a two-layered laminar flow system due to different diffusion of analytes. Wheeler et al. [86] developed an on-line sample preparation method for MALDI-MS, which depended on an electrowetting-on-dielectric-based technique. Ramsey and Collins [87] coupled SPE to micellar electrokinetic chromatography involving completely automated extraction, elution, injection, separation, and detection steps. The authors reported fast analysis of rhodamine B yielding preconcentration factors more than 200 in less than 5 min with 60 femtomolar as the detection limit. The schematic representation of this hyphenation is shown in Fig. 5.8 indicating sample preparation and separation parts clearly. Legendre et al. [88] described a chip-based online solid phase extraction for DNA and polymerase chain reaction in NLC. The amount injected was 600 nL of a blood sample. Xiao et al. [89] presented a sample preparation method on PDMS/glass chip coupled to gas chromatography. The authors tested this assembly for analysis of ephedrine from aqueous solution and reported good reproducibility of extraction and analysis.

Stachowiak et al. [90] described a fully automated system for the detection of an aerosolized bacterial biothreat agents, *Bacillus subtilis*, using a microfluidic sample preparation system chip coupled with gel electrophoresis. Furthermore, the authors investigated reducing response time, multiple microfluidic component modules, including aerosol collection via a commercially available collector, concentration, thermochemical lysis, size exclusion chromatography, and fluorescent labeling. Analysis can be completed within 10 min using this method. Soper et al. [91] reported a chip-based nanoreactor coupled to NCE for analysis of oligonucleotides. The nanoreactor is made of fused silica capillary tubes (10 to 20 cm $\times$ 20 to 50 μm i.d.) with fluid pumping via electroosmotic flow. The chip integration comprises an injector,

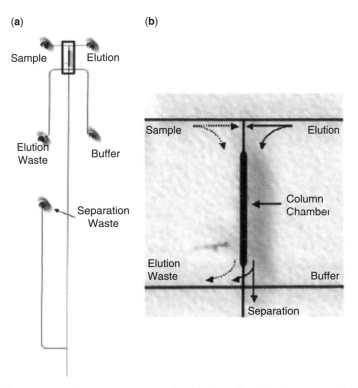

Figure 5.8   Schematic representation of a chip-based solid phase extraction-MEKC device. (a) Layout of the entire device and (b) expanded view of the extraction region of the device. The dotted lines represent the direction of fluid flow during extraction; the solid lines signify flow during elution/injection (Narrow channels are $\sim$55 μm wide, column chamber is $\sim$210 μm wide, with all channels $\sim$15 μm deep.) [87].

separation channel (length 6 cm, width 30 μm, depth 50 μm) and a dual fiber optic, near-infrared fluorescence detector. Xu et al. [92] reported online super-charging preconcentration of sodium dodecyl sulfate-protein complexes followed by their separation by microchip gel electrophoresis.

Dahlin et al. [93] described a PDMS chip-based solid phase extraction-nanocapillary electrophoresis-electrospray ionization-time-of-flight mass spectrometry (SPE-NCE-ESI-TOF-MS) assembly for the analysis of a mixture of six peptides. The chip was fabricated by molding PDMS in steel wires in a mold. When removed the wires give 50 μm cylindrical channels. The fused silica capillaries were inserted into the structure in a tight fit connection and the inner walls of the capillaries and the PDMS microchip channels were modified with a positively charged polymer (PolyE-323). Tuomikoski et al. [94] integrated a chip-based SPE coupled to NCE by fabricating multiple layers of SU-8 polymer. The nano-SPE-NCE device has a fluid reservoir of 15:1 high aspect ratio pillars with an electrophoresis channel 25 mm long. The authors simulated the device with Femlab software and fluorescein was used as the detecting analyte for the operational performance of the chip.

Huang et al. [95] developed online isotachophoretic preconcentration coupled to gel electrophoresis for separation of sodium dodecyl sulfate-proteins on a microchip. The schematic representation of this unit is shown in Fig. 5.9, which shows different parts of this device. Similarly,

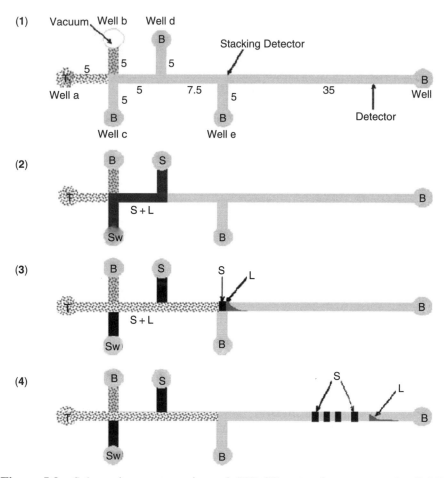

Figure 5.9 Schematic representation of ITP-GE procedure on a microfluidic device. S: sample, SW: sample waste, B: background electrolyte, L: leading electrolyte, T: terminating electrolyte. (1) B and T loading; (2) S and L injection—S at ground, SW at high voltage; (3) stacking—T at ground, B (well 6) at high voltage; (4) separation—B (well 5) at ground, B at high voltage [95].

Mohamadi et al. [96] reported online preconcentration of human serum albumin (HSA) and its immunocomplex with a monoclonal antibody on-chip coupled to isotachophoresis. The sample injection, preconcentration, and separation were carried out continuously and controlled by a sequential voltage switching program. Preconcentration was carried out with on-chip nondenaturing gel electrophoresis in methylcellulose solution. Furthermore, the authors applied this method for immunoassay of HSA. The separation of HSA and its immunocomplex was achieved in 25 seconds in 1 cm of the microchannel with induced fluorescence detection at 7.5 pM.

In the last few decades sample stacking was thought to be the best preconcentration technique in capillary electrophoresis, which has been tested in NCE [97]. Many modifications have been made in sample stacking, which include field-amplified sample stacking (FASS) [98–100], large-volume sample stacking (LVSS) [101,102], pH-mediated stacking [103,104], and micellar electrokinetic chromatography (MEKC) stacking [105–107]. Some reviews have been published on sample stacking techniques for a wide variety of compounds [97,108–117]. Terabe and coworkers [118–120] developed cation- and anion-selective exhaustive injection-sweeping-micellar electrokinetic chromatography (CSEI-, ASEI-sweeping-MEKC) methods for increased sensitivity and detection. Britz-McKibbin and colleagues [121,122] designed an online focusing method, based on different mobilities of cationic analytes between background electrolyte (BGE) and sample matrix, which is called a velocity difference-induced focusing (V-DIF) method.

Gong et al. [123] reported online sample preconcentration using field-amplified stacking injection in NCE. Pressure-driven flows into or from the branch channels, due to bulk velocity, can be used for liquid transportation in the channels. The detection sensitivity was improved by 94-, 108-, and 160-fold for fluorescein-5-isothiocyanate, fluorescein disodium, and 5-carboxyfluorescein, respectively, relative to a traditional method. Similarly, Zhang and Yin [124] developed a multi-T microchip integrated field-amplified sample stacking (FASS) coupled with NCE. The authors reported a volumetrically defined large sample plug was formed in one step within 5 seconds by negative pressure in the headspace of two sealed sample waste reservoirs. The authors reported precision in migration time and RSD as 3.3% and 1.3% for rhodamine123 (Rh123) and fluorescein sodium salt, respectively. Schematic representation of the channel design of a multi-T microfluidic chip and NCE system with negative-pressure, large-volume sample injection is shown in Fig. 5.10. Jung et al. [125] designed, fabricated, and characterized a novel field-amplified sample stacking (FASS)-NCE chip having photo-initiated porous polymer for the analysis of fluorescein and bodipy. Furthermore, the authors described 1000-fold signal increase during detection.

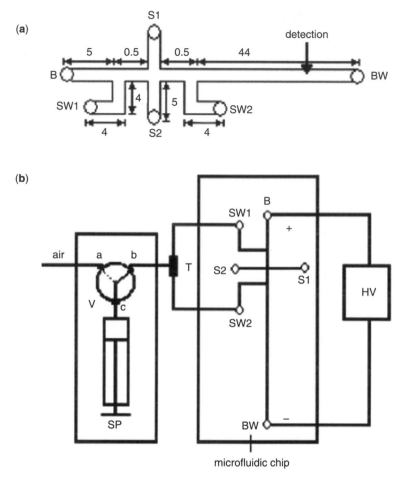

Figure 5.10 Schematic representation of (a) channel design of the multi-T micro-fluidic chip and (b) NCE with negative-pressure, large-volume sample injection. SP, syringe pump; V, 3-way valve; HV, high-voltage power supply; T, T-shaped connector [124].

This polymer material provided a region of high flow resistance, which allowed electromigration of sample ions resulting in preconcentration. Lichtenberg et al. [126] developed a microchip device for field amplification stacking (FAS), which allowed the formation of comparatively long, volume-trically defined sample plugs with a minimal NCE bias. The authors studied fluidic effects, which arose from solutions with mismatched ionic strengths, in chip-based electrokinetically. Furthermore, these authors developed a new chip layout for full column stacking with subsequent sample matrix removal by polarity switching.

Figures 5.11 and 5.12 [97] describe the working principles of online FASS and LVSS sample stacking methods. In FASS, the sample is prepared in a low conductivity matrix, which is different from BGE. The analytes move due to a proportionally greater electric field across the sample zone. Once the analytes reach the boundaries between the sample zone and BGE, the electric field strength suddenly drops causing poor migration, and resulting in focusing of analytes near the boundaries. Finally all analytes move toward detection window due to greater electroosmotic flow (EOF). In LVSS, experimental conditions are the same except the electrode polarity with reversed EOF. The sample is dissolved either in a low conductivity buffer or water. Samples with high conductivity may be stacked by a pH-mediated technique in which samples are prepared in a high ionic strength medium and electrokinetically injected into the capillary followed by the injection of strong acid electrophoretically and high voltage. Sample preconcentration in MEKC can be achieved in normal and reversed phase modes. The sample is dissolved in a

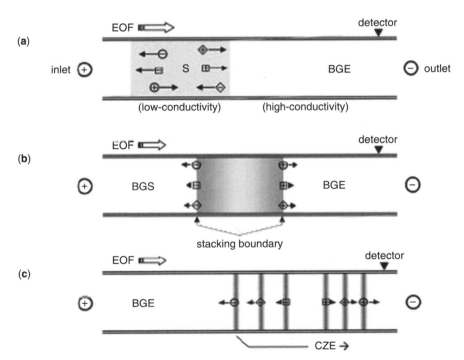

Figure 5.11 Schematic diagrams of the FASS model. (a) Capillary conditioned with a BGE, sample, injection and a high positive voltage, (b) focusing of analytes near the boundaries between the sample zone and BGE, and (c) stacked analytes separated [97].

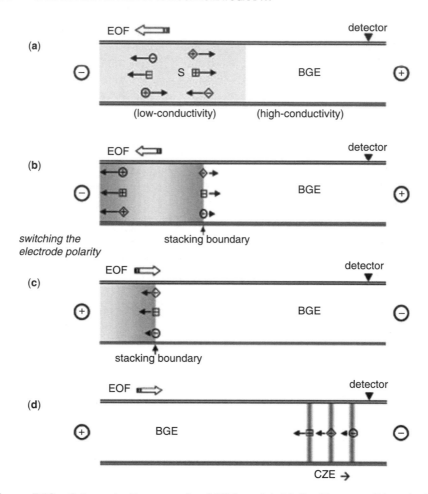

Figure 5.12 Schematic diagrams of an LVSS model. (a) Capillary conditioned with a BGE, sample injection, and a high positive voltage, (b) anionic analytes move toward the detection end and stack at one side of the boundary, (c) electrophoretic current carefully monitored reaching up to 95% to 99% of its original value, and (d) separation of analytes [97].

low conductivity buffer or water and BGE contains SDS to form the micelles. Sweeping sample stacking is an online sample concentration method for both charged and neutral analytes. The sample concentration effect depends on the pseudo-stationary phase, which sweeps the analytes. BGE contains a surfactant (SDS) to form a micellar buffer and the samples are dissolved in a non-micelle buffer. Normally, pH of the solution should be low to suppress EOF as the method is independent of EOF. At completion of injection of the sample solution, a negative polarity is applied to power NCE separation

TABLE 5.2 Sample Preconcentration Using Different Stacking Methods

Technique	Compounds	LOD	Concentration Efficiency	Refs.
FASS				
	Opioids	1.0 ng/mL	1000	[128]
	Alkaloid	0.7–0.9 ng/mL	1000	[129]
	Heroin metabolites	40 ng/mL	~10	[130]
	Pb, Hg, Se complex	<1 ng/mL	1700	[131]
LVSS				
	Phenolic species	5–25 ng/mL	40	[132]
	Food colorants	2–26 ng/mL	80	[133]
	Peptides	0.3–2 pM	10–15	[134]
	Anions	ppb levels	>300	[135,136]
	Bromide ion	15.0 ng/mL	—	[137]
pH-mediated				
	Haloacetic acid	0.5–4.0 ng/mL	97–120	[138]
Stacking-MEKC				
	Plant hormones	0.3 ng/mL	10–600	[139]
	Triazine	3.3–8.5 ng/mL	4–12	[140]
	Sildenafil citrate	17 ng/mL	11	[141]
	Herbicides	0.1–2.7 ng/mL	70–1540	[142]
	Riboflavin	20.0 ng/mL	1.8–6.2	[143]

(Continued)

TABLE 5.2 *Continued*

Technique	Compounds	LOD	Concentration Efficiency	Refs.
Sweeping	Steroids	1.7–9.6 ng/mL	1541–2564	[108]
	Proguanil, 4-chlorophenylbiguanide, & cycloguanil	10–20 ng/mL	—	[144]
	Estrogens	53–100 nM	—	[145]
	Doxorubicin & daunorubicin	1 nM	—	[146]
	trans-Resveratrol	5.0 ng/mL	1500	[147]
	Corticosterone	3.0 ng/mL	2000	[148]
	17-hydroxycorticosterone	3.0 ng/mL	2300	[148]
	Lysergic acid diethylamide	16.0 ng/mL	—	[149–151]
	Iso-lysergic acid diethylamide	22.0 ng/mL	—	[149–151]
	Lysergic acid *N,N*-methylpropamide	18.0 ng/mL	—	[149–151]
	Corticosterone	5.0 ng/mL	—	[152]
	17-Hydroxy corticosterone	4.0 ng/mL	—	[152]
	Bisphenol A & alkylphenols	6.5–21 ng/mL	41–69	[153]
	p-tert-butylphenol	24.0 ng/mL	122	[154]
	2,4,6-trichlorophenol	19.0 ng/mL	360	[154]
	Bisphenol A & alkylphenols	30–159.0 ng/mL	29–67	[155]
	NSAs	0.47–0.96.0 ng/mL	670–760	[156]

Steroids	9–20.0 ng/mL	270–370	[157]
2-Naphtholic acid & salicylic acid	0.4–31.0 ng/mL	590–600	[157]
Triazines	9–15.0 ng/mL	30–110	[158]
4-chlorophenol. 4-ethylphenol & 5-methylphenol	19–28.0 ng/mL	54–100	[159]
Heptanopheron, quinine	—	50	[160]
Quinine	42.0 ng/mL	580	[160]
	3–5.0 ng/mL	5800	[160]
Corticosteroids	50.0 ng/mL	—	[161]
Estrogens	118.0 ng/mL	—	[162]
Laulanosine & 1-naphthylamine	4.1–8.0 ppt	550,000	[118]
Aromatic amines	0.10 ppt	12,000	[163]
Paraquat, diquat & difenzoquat	0.075–1.0 ng/mL	3000	[164]
Corticosterone	5.0 ng/mL	1200	[148]
17-Hydroxy corticosterone	4.0 ng/mL	1500	[148]
Lysergic acid diethylamide	58 pg/mL	—	[149]
Dansyl amino acid	0.8–12 ng/mL	1150	[119]
Naphthalenedisulfonic acid	60–80 ppt	5800	[119]
Phenoxy acidic herbicides	100 ppt	100,000	[165]

Abbreviations: FASS, field-amplified sample stacking; LOD, limit of detection; LVSS, large-volume sample stacking; MEKC, micellar electrokinetic chromatography.

resulting in separation and good sensitivity. Zhong et al. [127] described a microchip-based online SPE column for analysis of λ-DNA. The authors reported application of this unit by purifying PCR-amplifiable genomic DNA from human hepatocellular carcinoma (HepG2) cells and human whole blood samples. Furthermore, the authors presented more than 80% recovery with 10% as RSD. The applications of these stacking methods for different analytes are given in Table 5.2.

5.12. CONCLUSION

Sample preparation in NLC and NCE is the most important step in analysis due to the nano nature of these modalities. The sampling should be carried out in such a way as to avoid changes in the chemical composition of the sample. The quantitative values of species depend on the strategy adopted in sample preparation. Extraction recoveries may vary from one species to another and they should, consequently, be assessed independently for each compound as well as for the compounds together. Materials with an integral analyte, that is, bound to the matrix in the same way as the unknown, which is preferably labeled (radioactive labeling) would be necessary, which is called method validation. As discussed above few papers described off- and online sample preparation methods on microfluidic devices. Of course, online methods are superior due to lower risk of contamination and error of methods. Not much work been carried out on online nanosample preparation devices, which need more research. Briefly, to get maximum extraction of analytes, sample preparation should be handled very carefully.

REFERENCES

1. R.E. Majors, *LC-GC*, **13**, 555 (1995).
2. P. Gebauer, W. Thormann, P. Bocek, *Electrophoresis*, **16**, 2039 (1995).
3. W.C. Brumley, *J. Chromatogr. Sci.*, **33**, 670 (1995).
4. Ph. Quevauviller, *J. Chromatogr. A*, **750**, 25 (1996).
5. L. Krivankova, P. Bocek, *J. Chromatogr. B*, **689**, 13 (1997).
6. M. Valcarcel, A. Rios, L. Arce, *Crit. Rev. Anal. Chem.*, **28**, 63 (1998).
7. L. Junting, C. Peng, O. Suzuki, *Forensic Sci. Int.*, **97**, 93 (1998).
8. P.R. Haddad, P. Doble, M. Macka, *J. Chromatogr. A*, **856**, 145 (1999).
9. W. Liu, H.K. Lee, *J. Chromatogr. A*, **834**, 45 (1999).
10. A.R. Timerbaev, W. Buchberger, *J. Chromatogr. A*, **834**, 117 (1999).

11. A.R. Timerbaev, *Talanta*, **52**, 573 (2000).
12. N.H. Snow, *J. Chromatogr. A*, **885**, 445 (2000).
13. S. Ulrich, *J. Chromatogr. A*, **902**, 167 (2000).
14. J.L. Gomez-Ariza, E. Morales, I. Giraldez, D. Sanchez-Rodas, A. Velasco, *J. Chromatogr. A*, **938**, 211 (2001).
15. J. Pawliszyn, *Adv. Exp. Med. Biol.*, **488**, 73 (2001).
16. A.R. Timerbaev, *Electrophoresis*, **23**, 3884 (2002).
17. G. Vas, K. Vekey, *J. Mass. Spectrom.*, **39**, 233 (2004).
18. G. Theodoridis, G.J. de Jong, *Adv. Chromatogr.*, **43**, 231 (2005).
19. F. Pragst, *Anal. Biochem.*, **388**, 1393 (2007).
20. I. Ali, H.Y. Aboul-Enein, *Chiral pollutants: Distribution, toxicity and analysis by chromatography and capillary electrophoresis*, Chichester: Wiley (2004).
21. I. Ali, H.Y. Aboul-Enein, *Instrumental methods in metal ions speciation: Chromatography, capillary electrophoresis and electrochemistry*, New York: Taylor & Francis (2006).
22. *Official Control of Foodstuffs Directives*, Council Directive 85/591/EEC, Off J L 372: 0050-0052 (1985).
23. *Official Control of Foodstuffs Directives*, Council Directive 87/524/EEC, Off J L 306: 0024-0031 (1987).
24. *Official Control of Foodstuffs Directives*, Council Directive 2001/22/EEC, Off J L 077: 0014-0021 (2001).
25. G.F. Evans, T.A. Lumplein, D.L. Smith, M.C. Somerville, *J. Air Waste Manag. Assoc.*, **42**, 1319 (1992).
26. J.A.C. Broekaert, S. Gücer, F. Adams (Eds.), *Metal speciation in the environment*, Berlin: Springer (1990).
27. M. Schweigkofler, R. Niessner, *Environ. Sci. Technol.*, **33**, 3680 (1999).
28. J.M. Boiano, M.E. Wallace, W.K. Sieber, J.H.J. Wang, K. Ashley, *J. Environ. Monit.*, **2**, 329 (2000).
29. K.S. Roinestad, J.B. Louis, J.D. Rosen, *J. AOAC Int.*, **76**, 1121 (1993).
30. R.C. Clark, M. Blumer, O.S. Raymond, *Deep Sea Res.*, **14**, 125 (1967).
31. H. Gaul, U. Ziebarth, *Dtsch. Hydrogr. Z*, **36**, 191 (1983).
32. T.C. Sauer, G.S. Durrell, J.S. Brown, D. Redford, P.D. Boehm, *Mar. Chem.*, **27**, 235 (1989).
33. A.D. Eaton, L.S. Clesceri, A.E. Greenberg (Eds.), *Standard methods for the examination of water and wastewater*, 19th Edition, Washington, DC: American Public Health Association (1995).
34. G. Kateman, *Chemometrics and sampling strategies*, Berlin: Springer (1987).
35. M.P. Guy, *Sampling of heterogeneous and dynamic material systems*, Amsterdam: Elsevier (1992).

36. J.W. Einax, H.W. Zwanziger, S. Geiss, *Chemometrics in environmental analysis*, VCH, Weinheim (1997).

37. H. Emons, J.D. Schladot, M.J. Schwuger, *Chemosphere*, **34**, 1875 (1997).

38. S. Weigel, K. Bester, H. Hühnerfuss, *J. Chromatogr. A*, **912**, 151 (2001).

39. L. Arce, A. Rios, M. Valcarcel, *J. Chromatogr. A*, **791**, 279 (1997).

40. L. Arce, A. Rios, M. Valcarcel, *Fres. J. Anal. Chem.*, **360**, 697 (1998).

41. T.A. Pressley, J.E. Longbottom, *USEPA Off. Res. Dev.*, [Rep] EPA-600/4-82-006: 31 (1982).

42. J. Faller, H. Hühnerfuss, W.A. König, R. Krebber, P. Ludwig, *Env. Sci. Technol.*, **25**, 676 (1991).

43. Z. Zhang, J. Pawliszyn, *Anal. Chem.*, **65**, 1843 (1993).

44. Z. Zhang, M. Yang, J. Pawliszyn, *Anal. Chem.*, **66**, 844A (1994).

45. S.B. Hawthorne, D.J. Miller, M.S. Krieger, *J. Chromatogr. Sci.*, **27**, 347 (1989).

46. A.B. Jemere, R.D. Oleschuk, F. Ouchen, F. Fajuyigbe, D.J. Harrison, *Electrophoresis*, **23**, 3537 (2002).

47. S. Ekström, J. Malmström, L. Wallman, M. Löfgren, J. Nilsson, T. Laurell, G. Marko-Varga, *Proteomics*, **2**, 413 (2002).

48. P. Jandik, B.H. Weigl, N. Kessler, J. Cheng, C.J. Morris, T. Schulte, N.J. Avdalovic, *J. Chromatogr. A*, **954**, 33 (2002).

49. P. Kuban, P.K. Dasgupta, K.A. Morris, *Anal. Chem.*, **74**, 5667 (2002).

50. Y.C. Xu, B. Vaidya, A.B. Patel, S.M. Ford, R.L. McCarley, S.A. Soper, *Anal. Chem.*, **75**, 2975 (2003).

51. M.C. Breadmore, K.A. Wolfe, I.G. Arcibal, W.K. Leung, D. Dickson, B.C. Giordano, M.E. Power, J.P. Ferrance, S.H. Feldman, P.M. Norris, J.P. Landers, *Anal. Chem.*, **75**, 1880 (2003).

52. B.S. Broyles, S.C. Jacobson, J.M. Ramsey, *Anal. Chem.*, **75**, 2761 (2003).

53. M. Gustafsson, D. Hirschberg, C. Palmberg, H. Jornvall, T. Bergman, *Anal. Chem.*, **76**, 345 (2004).

54. O. Hofmann, K. Murray, A.S. Wilkinson, T. Cox, A. Manz, *Lab. Chip*, **5**, 374 (2005).

55. R. Bozalongo, J.D. Carrillo, M.A. Torroba, M.T. Tena, *J. Chromatogr. A*, **1173**, 10 (2007).

56. T.N. Chiesl, W. Shi, A.E. Barron, *Anal. Chem.*, **77**, 772 (2005).

57. A. Bhattacharyya, C.M. Klapperich, *Anal. Chem.*, **78**, 788 (2006).

58. B.H. Huynh, B.A. Fogarty, P. Nandi, S.M. Lunte, *J. Pharm. Biomed. Anal.*, **42**, 529 (2006).

59. M.J. Shiddiky, Y.B. Shim, *Anal. Chem.*, **79**, 3724 (2007).

60. M. Buratti, O. Pellegrino, C. Valla, F.M. Rubino, C. Verduci, A. Colombi, *Biomed. Chromatogr.*, **20**, 971 (2006).

61. H. Hisamoto, S. Takeda, S. Terabe, *Anal. Bioanal. Chem.*, **386**, 733 (2006).

62. N. Lion, J.O. Gallon, H. Jensen, H.H. Girault, *J. Chromatogr. A*, **1003**, 11 (2003).

63. B.H. Timmer, K.M. van Delft, W. Olthuis, P. Bergveld, A. van den Berg, *Sensors & Actuators B*, **91**, 342 (2003).

64. S. Song, A.K. Singh, B. Kirby, *Anal. Chem.*, **76**, 4589 (2004).

65. R.S. Foote, J. Khandurina, S.C. Jacobson, J.M. Ramsey, *Anal. Chem.*, **77**, 57 (2005).

66. X.Y. Wang, C. Saridara, S. Mitra, *Anal. Chim. Acta*, **543**, 92 (2005).

67. L. Nyholm, *Analyst*, **130**, 599 (2005).

68. Y. Zhang, A.T. Timperman, *Analyst*, **128**, 537 (2003).

69. P.K. Wong, C.Y. Chen, T.H. Wang, C.M. Ho, *Anal. Chem.*, **76**, 6908 (2004).

70. Y.C. Wang, A.L. Stevens, J. Han, *Anal. Chem.*, **77**, 4293 (2005).

71. R. Dhopeshwarkar, S.A. Li, R.M. Crooks, *Lab. Chip*, **5**, 1148 (2005).

72. F.C. Leinweber, J.C.T. Eijkel, J.G Bomer, A. Van den Berg, *Anal. Chem.*, **78**, 1425 (2006).

73. Y. Li, D.L. DeVoe, C.S. Lee, *Electrophoresis*, **24**, 193 (2003).

74. B. Grass, R. Hergenroder, A. Neyer, D.J. Siepe, *Separation Sci.*, **25**, 135 (2002).

75. Y. Xu, C.X. Zhang, D. Janasek, A. Manz, *Lab. Chip*, **3**, 224 (2003).

76. D. Ross, L.E. Locascio, *Anal. Chem.*, **74**, 2556 (2002).

77. M.A. Holden, S. Kumar, E.T. Castellana, A. Beskok, P.S. Cremer, *Sensors & Actuators B*, **92**, 199 (2003).

78. Y.J. Liu, R.S. Foote, S.C. Jacobson, J.M. Ramsey, *Lab. Chip*, **5**, 457 (2005).

79. H. Chen, Q. Fang, X.F. Yin, Z.L. Fang, *Lab. Chip*, **5**, 719 (2005).

80. B.H. Huynh, B.A. Fogarty, R.S. Martin, S.M. Lunte, *Anal. Chem.*, **76**, 6440 (2004).

81. B.H. Huynh, B.A. Fogarty, P. Nandi, S.M. Lunte, *J. Pharm. Biomed. Anal.*, **42**, 529 (2006).

82. A.S. Yazdi, Z. Es'haghi, *J. Chromatogr. A.*, **1082**, 136 (2005).

83. Z. Long, D. Liu, N. Ye, J. Qin, B. Lin, *Electrophoresis*, **27**, 4927 (2006).

84. Z. Long, Z. Shen, D. Wu, J. Qin, B. Lin, *Lab. Chip*, **7**, 1819 (2007).

85. D.J. Wilson, L. Konermann, *Anal. Chem.*, **77**, 6887 (2005).

86. A.R. Wheeler, H. Moon, C.A. Bird, R.R.O. Loo, C.J. Kim, J.A. Loo, R.L. Garrell, *Anal. Chem.*, **77**, 534 (2005).

87. J.D. Ramsey, G.E. Collins, *Anal. Chem.*, **77**, 6664 (2005).

88. L.A. Legendre, J.M. Bienvenue, M.G. Roper, J.P. Ferrance, J.P. Landers, *Anal. Chem.*, **78**, 1444 (2006).

89. H. Xiao, D. Liang, G. Liu, M. Guo, W. Xing, J. Cheng, *Lab. Chip*, **6**, 1067 (2006).

90. J.C. Stachowiak, E.E. Shugard, B.P. Mosier, R.F. Renzi, P.F. Caton, S.M. Ferko, J.L. Van de Vreugde, D.D. Yee, B.L. Haroldsen, V.A.V. Noot, *Anal. Chem.*, **79**, 5763 (2007).

91. S.A. Soper, S.M. Ford, Y. Xu, S. Qi, S. McWhorter, S. Lassiter, D. Patterson, R.C. Bruch, *J. Chromatogr. A*, **853**, 107 (1999).

92. Z.Q. Xu, T. Ando, T. Nishine, A. Arai, T. Hirokawa, *Electrophoresis*, **24**, 3821 (2003).

93. A.P. Dahlin, S.K. Bergström, P. Andrén, K.E. Markides, J. Bergquist, *Anal. Chem.*, **77**, 5356 (2005).

94. S. Tuomikoski, N. Virkkala, S. Rovio, A. Hokkanen, H. Sirén, S. Franssila, *J. Chromatogr. A.*, **1111**, 258 (2006).

95. H. Huang, F. Xu, Z. Dai, B. Lin, *Electrophoresis*, **26**, 2254 (2005).

96. M.R. Mohamadi, N. Kaji, M. Tokeshi, Y. Baba, *Anal. Chem.*, **79**, 3667 (2007).

97. C.H. Lin, T. Kaneta, *Electrophoresis*, **25**, 4058 (2004).

98. D.S. Burgi, R.L. Chien, *Anal. Chem.*, **63**, 2042 (1991).

99. Z. Liu, P. Sam, S.R. Sirimanne, P.C. McClure, J. Grainger, D.G. Patterson, *J. Chromatogr. A*, **673**, 125 (1994).

100. S. Locke, D. Figeys, *Anal. Chem.*, **72**, 2684 (2000).

101. R.L. Chien, D.S. Burgi, *Anal. Chem.*, **64**, 1046 (1992).

102. D.S. Burgi, R.L. Chien, *Anal. Biochem.*, **202**, 306 (1992).

103. M.E. Hadwiger, S.R. Torchia, S. Park, M.E. Biggin, C.E. Lunte, *J. Chromatogr. B*, **681**, 241 (1996).

104. S. Park, C.E. Lunte, *J. Microcol. Sep.*, **10**, 511 (1998).

105. J.P. Quirino, S. Terabe, *J. Chromatogr. A*, **781**, 119 (1997).

106. C.X. Zhang, W. Thormann, *Anal. Chem.*, **70**, 540 (1998).

107. J.B. Kim, S. Terabe, *J. Chromatogr. A*, **979**, 131 (2002).

108. J.P. Quirino, S. Terabe, *Science*, **282**, 465 (1998).

109. J.P. Quirino, S. Terabe, *Anal. Chem.*, **71**, 1638 (1999).

110. J.L. Beckers, P. Bocek, *Electrophoresis*, **14**, 2747 (2000).

111. Z.K. Shihabi, *J. Chromatogr. A*, **902**, 107 (2000).

112. D.M. Osbourn, D.J. Weiss, C.E. Lunte, *Electrophoresis*, **21**, 2768 (2000).

113. J.P. Quirino, S. Terabe, *J. Chromatogr. A*, **902**, 119 (2000).

114. M.C. Breadmore, P.R. Haddad, *Electrophoresis*, **22**, 2464 (2001).

115. S. Sentellas, L. Puignou, M.T. Galceran, *J. Separation Sci.*, **25**, 975 (2002).

116. P. Britz-McKibbin, S. Terabe, *J. Chromatogr. A*, **1000**, 917 (2003).

117. J.P. Quirino, J.B. Kim, S. Terabe, *J. Chromatogr. A*, **965**, 357 (2002).

118. J.P. Quirino, S. Terabe, *Anal. Chem.*, **72**, 1023 (2000).

119. J.B. Kim, K. Otsuka, S. Terabe, *J. Chromatogr. A*, **932**, 129 (2001).

120. K. Isoo, S. Terabe, *Anal. Chem.*, **75**, 6789 (2003).

121. P. Britz-McKibbin, G.M. Bebault, D.D.Y. Chen, *Anal. Chem.*, **72**, 1729 (2000).

122. P. Britz-McKibbin, D.D. Chen, *Anal. Chem.*, **72**, 1242 (2000).

123. M. Gong, K.R. Wehmeyer, P.A. Limbach, F. Arias, W.R. Heineman, *Anal Chem.*, **78**, 3730 (2006).

124. L. Zhang, X.F. Yin, *J. Chromatogr. A*, **1137**, 243 (2006).

125. B. Jung, R. Bharadwaj, J.G. Santiago, *Electrophoresis*, **24**, 3476 (2003).

126. J. Lichtenberg, E. Verpoorte, N.F. Rooij, *Electrophoresis*, **22**, 258 (2001).

127. R. Zhong, D. Liu, L. Yu, N. Ye, Z. Dai, J. Qin, B. Lin, *Electrophoresis*, **28**, 2920 (2007).

128. A.B. Wey, W. Thormann, *J. Chromatogr. A*, **924**, 507 (2001).

129. S. Liu, Q. Li, X. Chen, Z. Hu, *Electrophoresis*, **23**, 3392 (2002).

130. A. Alnajjar, B. McCord, *J. Pharm. Biomed.*, **33**, 463 (2003).

131. W. Liu, H.K. Lee, *Electrophoresis*, **20**, 2475 (1999).

132. J. Kruaysawat, P.J. Marriott, J. Hughes, C. Trenerry, *Electrophoresis*, **24**, 2180 (2003).

133. H.Y. Huang, C.W. Chiu, S.L. Sue, C.F. Cheng, *J. Chromatogr. A*, **995**, 29 (2003).

134. N. Siri, P. Riolet, C. Bayle, F. Couderc, *J. Chromatogr. B*, **793**, 151 (2003).

135. J.P. Quirino, S. Terabe, *J. Chromatogr. A*, **850**, 339 (1999).

136. Y. He, H.K. Lee, *Anal. Chem.*, **71**, 995 (1999).

137. P. Rantakokko, T. Nissinen, T. Vartiainen, *J. Chromatogr. A*, **839**, 217 (1999).

138. C. Tu, L. Zhu, C.H. Ang, H.K. Lee, *Electrophoresis*, **24**, 2188 (2003).

139. B.F. Liu, X.H. Zhong, Y.T. Lu, *J. Chromatogr. A*, **945**, 257 (2002).

140. E. Turiel, P. Fernández, C. Pérez-Conde, C. Cámara, *Analyst*, **125**, 1725 (2000).

141. J.J.B. Nevado, J.R. Flores, G.C. Peñalvo, N.R. Fariñas, *J. Chromatogr. A*, **953**, 279 (2002).

142. R. Carabias-Martínez, E. Rodríguez-Gonzalo, P. Revilla-Ruiz, J. Domínguez-Alvarez, *J. Chromatogr. A*, **990**, 291 (2003).

143. A.K. Su, C.H. Lin, *J. Chromatogr. B*, **785**, 39 (2003).

144. R.B. Taylor, R.G. Reid, A.S. Low, *J. Chromatogr. A*, **916**, 201 (2001).

145. H. Harino, S. Tsunoi, T. Sato, M. Tanaka, *Fres. J. Anal. Chem.*, **369**, 546 (2001).

146. A. Gavenda, J. Sevcik, J. Psotova, P. Bednar, P. Bartak, P. Adamovsky, V. Simanek, *Electrophoresis*, **22**, 2782 (2001).

147. M.C. Hsieh, C.H. Lin, *Electrophoresis*, **25**, 677 (2004).

148. M.C. Chen, S.H. Chou, C.H. Lin, *J. Chromatogr. B*, **801**, 347 (2004).

149. C. Fang, J.T. Liu, S.H. Chou, C.H. Lin, *Electrophoresis*, **24**, 1031 (2003).

150. C. Fang, J.T. Liu, C.H. Lin, *J. Chromatogr. B*, **775**, 37 (2002).

151. C. Fang, J.T. Liu, C.H. Lin, *Talanta*, **58**, 691 (2002).

152. C.H. Wu, M.C. Chen, A.K. Su, P.Y. Shu, S.H. Chou, C.H. Lin, *J. Chromatogr. B*, **785**, 317 (2003).

153. S. Takeda, A. Omura, K. Chayama, T. Tsuji, K. Fukushi, M. Yamane, S.I. Wakida, S. Tsubota, S. Terabe, *J. Chromatogr. A*, **1014**, 103 (2003).

154. M.R. Monton, K. Otsuka, S. Terabe, *J. Chromatogr. A*, **985**, 435 (2003).

155. S. Takeda, A. Omura, K. Chayama, H. Tsuji, K. Fukushi, M. Yamane, S.I. Wakida, S. Tsubota, S. Terabe, *J. Chromatogr. A*, **979**, 425 (2002).

156. J.B. Kim, J.P. Quirino, K. Otsuka, S. Terabe, *J. Chromatogr. A*, **916**, 123 (2001).

157. J.B. Kim, K. Otsuka, S. Terabe, *J. Chromatogr. A*, **912**, 343 (2001).

158. C.E. Lin, Y.C. Liu, T.Y. Yang, T.Z. Wang, C.C. Yang, *J. Chromatogr. A*, **916**, 239 (2001).

159. M.R.N. Monton, J.P. Quirino, K. Otsuka, S. Terabe, *J. Chromatogr. A*, **939**, 99 (2001).

160. W. Shi, C.P. Palmer, *J. Separation Sci.*, **25**, 215 (2002).

161. J. Palmer, N.J. Munro, J.P. Landers, *Anal. Chem.*, **71**, 1679 (1999).

162. N.J. Munro, J. Palmer, A.M. Stalcup, J.P. Landers, *J. Chromatogr. B*, **731**, 369 (1999).

163. J.P. Quirino, Y. Iwai, K. Otsuka, S. Terabe, *Electrophoresis*, **21**, 2899 (2000).

164. O. Núñez, J.B. Kim, E. Moyano, M.T. Galceran, S. Terabe, *J. Chromatogr. A*, **961**, 59 (2002).

165. L. Zhu, C. Tu, H.K. Lee, *Anal. Chem.*, **74**, 5820 (2002).

CHAPTER 6

NANO-HIGH PERFORMANCE LIQUID CHROMATOGRAPHY

6.1. INTRODUCTION

Basically, high performance liquid chromatography (HPLC) is a modality of liquid chromatography having high performance in separations. It is often erroneously called high pressure liquid chromatography as it works at high pressure of the mobile phase. HPLC has emerged as a technique of choice in analytical science since 1980. It has been used widely for the analyses of almost all classes of compounds, including organic and inorganic moieties. A good reputation of HPLC is due to its high speed, sensitivity, reproducibility, and wide range of applications. The development of a variety of columns and coupling of various sample preparation units and detectors make it a superb analytical technique. Moreover, the use of aqueous, nonaqueous, and organic mobile phases enhances its application greatly. Various modifications have been made in HPLC from time to time, based on its working principles, including such techniques as normal phase, reversed phase, size exclusion, gel permeation micellar electrokinetic and capillary electrochromatography. Similarly, modifications in HPLC have been reported for the analyses of low amounts of sample or samples having poor concentration of ingredients and these are termed conventional, micro-, and nano-HPLC. Some aspects of analysis, such as instrumentation, sample preparation, and

Nanochromatography and Nanocapillary Electrophoresis. By Ali, Aboul-Enein, and Gupta
Copyright © 2009 John Wiley & Sons, Inc.

detection in nano-HPLC have already been discussed in earlier chapters; this chapter is focused on the applications of nano-HPLC for different molecules. Nano-HPLC is a popular technique in proteomics, pharmaceutical, and environmental analyses due to the inherent characteristic of low sample volumes or samples with poor ingredients. Some reviews have appeared in the literature dealing with applications for nano-HPLC [1–8].

6.2. NANO-HPLC

Nowadays, an interest is growing in analyzing nano amounts of samples in the shortest time and integrating both sample pretreatment and analytical separation, which has been realized in nano-HPLC. Therefore, nano-HPLC is gaining importance. Briefly, nano-HPLC is the latest trend in separation science globally. This chapter discusses the state of the art of nano-HPLC analyses in biological and environmental matrices. The use of precolumns is an important in nano-HPLC due to the small size of the separation channel and to avoid blockage of the channel. Reversed phase silica (C_2, C_8, and C_{18})-based microchips have been used in nano-HPLC [9]. Glass, silicon, and plastic chips have been success-fully applied in nano-HPLC. But there are some problems with this technique, such as design of the chip, introduction of stationary phase into the narrow chan-nels, and sealing of the microchannels. Monolithic silica (particle fixed, polymer based and molecular imprinted) has also been used in nano-HPLC, which results in good separation of a variety of molecules [10,11]. These were coated in the inner surface of the channel; *in situ* polymerization of continuous beds [12–15]. Presently, some microfluidic nano-HPLC chips are being commercia-lized integrating the column, connection, and nanospray emitters.

6.3. APPLICATIONS

Nano-HPLC has been used for the separation and identification of molecules in different matrices such as proteomic, pharmaceutical, and environmental sciences. Of course, sample preparation is the most important task prior to nano-HPLC analysis, which has already been discussed in this book. The applications of nano-HPLC in different matrices are discussed in the following sections.

6.3.1. Nano-HPLC of Biological Matrices

Nano-HPLC can be used for the analysis of any compound in any sample of a biological nature. The selection of the chip coating materials and mobile phase depends on the type of molecules to be analyzed. Applications of nano-HPLC in biological samples are discussed in the following sections.

6.3.1.1. Proteomics

Proteomics (peptide mapping, protein sequencing, determination of post-translational modification of proteins) is one of the newly emerged areas in which nano-HPLC is a very useful technique because of the low amounts of proteins present in living systems. nano-HPLC has emerged as one of the useful tools in proteomics. Some workers have used this modality of liquid chromatography in proteomics study; some of them are discussed here. Licklider et al. [16] developed a chip-based liquid chromatography and mass spectrometry electrospray source on a silicon chip and used it for automated MS/MS analysis of a mixture of tryptic peptides. The authors advocated an efficient electrospray interface to mass spectrometry on-chip structures. Ivanov et al. [17] used 20 μm i.d. polymeric polystyrene divinylbenzene (PS-DVB) monolithic chip columns for analysis of tryptic digest peptide mixtures with high sensitivity. Mao et al. [18] described lectin affinity chromatography on a miniaturized microfluidic device with a lectin affinity monolith column prepared in the microchannel of a microfluidic chip. Three glycoproteins, turkey ovalbumin, chicken ovalbumin, and ovomucoid were separated successfully by lectin affinity chromatography. This integrated system reduces running time from 4 hours to 400 seconds. The detection limit achieved was in the picogram range.

Song et al. [19] described determination of neurotransmitter histamine in rat brain tissues by Nano-HPLC-MS. The separation channel was 75 μm id packed with C_{18} particles of 5 μm size. The mobile phase used was water, acetonitrile, formic acid with 1-heptanesulfonic acid as ion-pairing reagent. Bedair and El Rassi [20] used nano-HPLC for the analyses of glycoproteins and glycans including their isolation and pre-concentration from diluted samples. Okanda and El Rassi [21] reported separation and identification of glycoproteins using nano-HPLC microdevices. Yin et al. [22] reported separation of bovine serum albumin (BSA) proteins by nano-HPLC-MS. Generally, two-dimensional nano-HPLC was commonly performed using ion exchange chromatography to create fractions of peptides, followed by reversed phase chromatographic analysis of each of these fractions. It is important to mention that special precautions should be taken to keep separated components in the first column [23,24].

Ghitun et al. [25] used nano-HPLC for the analysis of complex protein digests by integrating a 30 nL precolumn (trap) and a 10 μm i.d. nano-electrospray emitter coupled with a TOF-MS. The separation channel was of $45 \times 0.075 \times 0.050$ mm dimension packed with either Zorbax C_{18} or C_3 of 2.1, 3.5, or 5 μm particle sizes. The complexity of the samples and dynamic range of the peptide abundance was overcome by adopting two-dimensional nano-HPLC methods. Huang et al. [26] described a label-free nano-HPLC-MS to investigate the proteomic profile of cerebrospinal fluid (CSF) by addressing quality control, sample replication steps, and the adaptation of pattern recognition methods for the detection of experimental variation and (most importantly) putative biomarkers. The authors tested 20 CSF samples (10 samples from healthy volunteers and 10 from patients with schizophrenia). Yue et al. [27] described immobilized metal affinity chromatography (IMAC) on an integrated glass microchannel for the analyses of phosphopeptide fragments from β-casein in peptide mixtures.

Wang et al. [28] described protein analyses of serum samples from 99 ovarian cancer patients, 87 healthy volunteers, and 21 patients with other ovarian diseases by using metal affinity chromatography protein chip IMAC3 and cation exchange protein chip WCX2. The detection was achieved by time-of-flight mass spectrometry. Ishida et al. [29] described a microchip for reversed phase nano-HPLC using porous monolithic silica with a double T-shaped injector and an approximately 40 cm serpentine separation channel. Octadecyl modified monolithic silica was coated in the channel using the sol gel process. The authors also studied the effect of geometry of turn sections on band dispersion under pressure-driven flow. The separation efficiencies of 15,000 to 18,000 plates/m for catechins were reported in this work. This system was used to analyze catechins by using methanol-water (3 : 2, v/v) and 0.02 molar phosphate buffer (pH 2.0) at 300 nL/min flow rate.

Srbek et al. [30] used nano-HPLC-ESI-MS for the separation and identification of some biological protein mixtures in rice seed protein sample. The authors used a 1100 capillary pump of Agilent Technologies with 4 μL/min as flow rate of 0.1% formic acid for sample preconcentration on chip. Furthermore, a 1100 nano-HPLC pump (Agilent Technologies) with 300 nL/min. flow rate of water-acetonitrile with 0.1% formic acid was used as mobile phase for separation and identification of protein samples. Nano-chromatographic separations of mixtures of rice, lentil, and human nail proteins are shown in Fig. 6.1. Gaspari et al. [31] described protein identification at the nanogram level using a nano-HPLC-MS system. The authors used an off-line two-step sample clean up subsequent to a protein digestion method and the purified samples were loaded onto a nano-HPLC-MS device. The mobile phase used was water-acetonitrile-trifluoroacetic acid (97.95 : 2 : 0.05, v/v/v). The selected ion chromatogram of BSA ions is shown in Fig. 6.2 indicating a good resolution of components. Lanckmans et al. [32] determined extracellular concentration of angiotensin IV (Ang IV) peptide in rat brain by using nano-HPLC-MS-MS and the detection limit

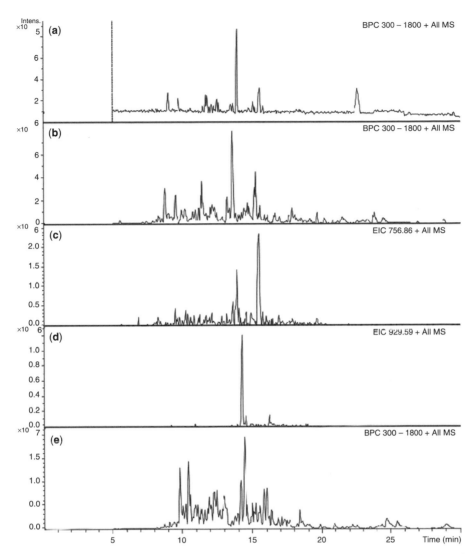

Figure 6.1   The base peak chromatograms of mixtures of rice, lentil, and human nail proteins containing 4.0 (A), 40 (B), and 320 ng (E) of each sample [30].

obtained was 50 pM. The mobile phase used was water-acetonitrile-formic acid $(98:2:0.1 \, \text{v/v/v})$ at $300 \, \text{nL/min}$ flow rate. The authors also advocated nano-HPLC-MS-MS for the determination of met-enkephalin and neurotensin in dialysates from rat.

Vollmer et al. [33] studied human nucleolus proteome using off-line strong cation exchange chromatography and microfraction collection combined with

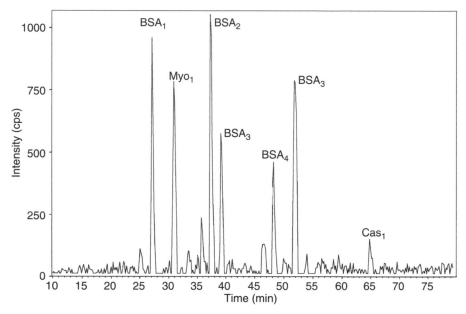

Figure 6.2   The chromatogram of ions at m/z 449.8 (BSA_1), 653.4 (BSA_2), 820.5 (BSA_3), 710.4 (BSA_4), 751.9 (Myo_1), 692.9 (Cas_1) belonging to bovine serum albumin, myoglobin, and α-casein digests (5 ng each) [31].

nano-HPLC-MS. The authors reported 206 unique proteins were identified in the International Protein Index human database corresponding to 2,024 unique tryptic peptides. Staes et al. [34] described a special type of nano-HPLC holding larger enrichment columns for gel-free proteome studies. A tryptic digest of a human T-cell proteome was fractionated by strong cation exchange chromatography and selected fractions were analyzed by MS/MS on an IT mass spectrometer using nano-HPLC-MS-MS. The authors reported their nano-HPLC-MS-MS system capable of separating very complex peptide mixtures. Hardouin et al. [35] described nano-HPLC-MS for analyses of proteins obtained from two-dimensional gel electrophoresis or chromatography. The retention times and m/z were found to be reproducible for identifying peptide sequences without ambiguity. It has been advocated by the authors that the device is a valuable tool in biomarker discovery programs, particularly for identifying low-abundance proteins. Furthermore, the same group [36] described a nano-HPLC device interfaced to an MS-MS detector for identification of autoantigens. The authors described this system as highly reproducible and efficient. Li et al. [37] reported an integrated and modular microsystem for rapid analyses of trace-level tryptic digests for proteomics applications. This microsystem included an autosampler, a large channel (2.4 μL total volume) and separation channels, together with a low dead volume enabling the interface to nanoelectrospray mass spectrometry. The separation channel contained C_{18} reversed phase silica packing. The authors reported the separation of peptides and tryptic digests with a throughput of up to 12 samples per hour with a detection limit of 5 nM.

6.3.1.2. Drugs Development and Design   Nano-HPLC has been used for both drug discovery and development programs by studying drug degradation, metabolism, and quality control [38]. Wistuba and Schurig [39] and Lanckmans et al. [40] used nano-HPLC for quantitative analysis of oxcarbazepine and its active metabolite 10,11-dihydro-10-hydroxycarbamazepine with internal standard of 2-methyl-5H-dibenz(b,f)azepine-5-carboxamide in rat brain microdialysates. Chmela et al. [41] developed a miniaturized hydrodynamic chromatography chip (1 μm deep and 1000 μm wide, integrated with a 300 pL injector on a silicon substrate) and tested for the separation of fluorescent nanospheres, macromolecules, synthetic polymers, biopolymers, and particles. The authors reported this system as fast, efficient, and low solvent consumption. The silicon microtechnology provided precisely defined geometry, high rigidity, and compatibility with organic solvents or high temperature. The separations obtained in 3.0 minutes showed high performance of the device. Furthermore, the same group [42] designed a planar chip with a 1 μm high, 0.5 mm wide, and 69 mm long channel having an integrated 150 pL injection structure and a 30 μm deep and 30 μm wide detection cell, suitable for UV absorption detection. The authors reported a hydrodynamic separation within 70 seconds of several synthetic polymers, biopolymers, and particles. Song et al. [19] reported the analysis of neurotransmitter histamine in alcoholic beverages by nano-HPLC-MS.

Shih et al. [43] described integrated microelectromechanical system (MEMS)-HPLC consisting of a parylene HPLC column, an electrochemical sensor, a resistive heater, and a thermal-isolation structure for separation of a mixture of derivatized amino acids. The separation column was 8 mm long, 100 μm wide, 25 μm high, packed with 5 μm C_{18} coated beads. The authors used a temperature gradient scanning from 25°C to 65°C with a rise rate of 3.6°C per minute. Fanali et al. [44] used nano-LC-MS for the analysis of nadolol, oxprenolol, alprenolol, and propranolol using terbutaline as the internal standard. The authors used C_{18} as stationary phase of 5 μm particle size packed in a 75 μm i.d. capillary connected to a MS via a sheath-liquid interface or a sheathless nanospray interface. The mobile phase used was 500 mM ammonium acetate (pH 4.5)-water-methanol (1 : 8 : 91, v/v) and analysis was completed in less than 10 minutes with 0.1 ng/mL limit of detection. The method was used for analysis of oxprenolol in pharmaceutical preparations. Fuentes and Woolley [45] described electrolysis-based micro-pumps integrated microfluidic channels in micromachined glass substrates. The authors developed a stationary phase by coating the microchannel with 10% (w/w) chlorodimethyloctadecylsilane in toluene. The authors also reported a chip electrochemical pumping enabled the loading of picoliter samples with no dead volume between injection and separation. The mobile phase used was acetonitrile-50 mM acetate buffer (pH 5.45) (70 : 30, v/v) for the separation of three fluorescently labeled amino acids in 40 seconds with an efficiency of 3000 theoretical plates in a 2.5 cm long channel.

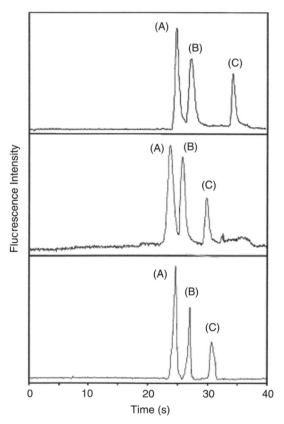

Figure 6.3   The chromatograms of a mixture of 0.5 μm MFITC-derivatized amino acids in a 2.5 cm × 100 μm × 5 μm LC microchip column. (A) Aspartic acid, (B) glycine, and (C) phenylalanine; mobile phase: acetonitrile-50 mM acetate (pH 5.45) (70 : 30, v/v) [45].

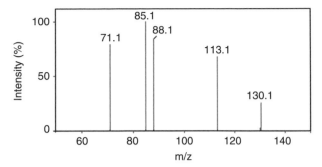

Figure 6.4   ESI-MS-MS spectrum of metformin [46].

The chromatographic separation of a mixture of 0.5 μm fluorescein-5-isothio-cyanate (FITC)-derivatized amino acids on a 2.5 cm × 100 μm × 5 μm LC microchip column is shown in Fig. 6.3, indicating a sharp resolution within 40 seconds.

Lu and Feng [46] described a nano-HPLC method with tandem mass spectrometry (nano-HPLC-MS-MS) for the determination of basic compounds (biguanides, metformin, and phenformin) in plasma. The authors reported plasma handling by deproteination and acidification with heptafluorobutyric acid (HFBA). The resulting supernatant was injected onto a nano-HPLC-MS-MS system. The chromatographic conditions were C_{18} column with methanol-40 mM aqueous HFBA (75 : 25, v/v) as mobile phase at a flow rate of 800 nL/min. The nano-HPLC-MS-MS chromatograms of metformin and phenformin are shown in Fig. 6.4 and Fig. 6.5, respectively, indicating good separation in short analysis times. Furthermore, the authors studied the influence of methanol concentrations (varied from 40% to 80%) in mobile phase on the retention time of these molecules (Fig. 6.6). The authors demonstrated the separation efficiency of 1.0 cm columns with different inner

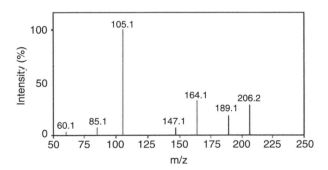

Figure 6.5   ESI-MS-MS spectrum of phenformin [46].

diameters (75, 100, and 150 μm) by taking metformin and phenformin (1 mg/mL) as model compounds. It was found that the retention times of metformin were almost the same for columns with 75 and 100 μm inner diameters (Fig. 6.6a). But for 150 μm inner diameter the retention time of metformin was increased, as the aqueous content of the mobile phase was up to 50% (Fig. 6.6b). Furthermore, the authors observed that the retention time of phenformin increased as the aqueous content of the mobile phase was up to 50% (Fig. 6.6a) for 75 and 100 μm inner diameter columns. But for a column of 150 μm inner diameter, the maximum retention observed was below 30%

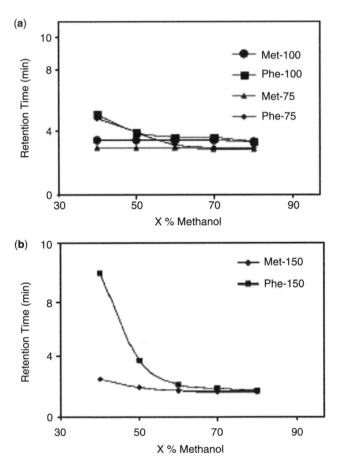

Figure 6.6 Effect of methanol concentration in the mobile phase on the retention time of metformin and phenformin using varied concentrations of methanol with 40 mM HFBA [46].

methanol for phenformin (Fig. 6.6b). The authors have explained this behavior as due to more abundant stationary phase in the 150 μm inner diameter column in comparison to 75 and 100 μm inner diameter columns and, hence, the former retained these molecules for longer. In addition, the authors also gave the explanation for this sort of chromatographic retention based on the more hydrophilic nature of metformin than phenformin with poor retaining of the former molecule in hydrophobic columns.

6.3.1.3. Environmental Analysis During the last few years interest in nano-HPLC for use in environmental samples has grown due to the presence of low amounts of pollutants in these samples. Nano-HPLC is an important tool in environmental monitoring in industrial areas where many assays must be performed to ensure health and public safety. There is very little in the literature in the application of nano-HPLC in environmental analysis [47]. Cappiello et al. [48] used nano-HPLC coupled to MS for the analyses of pesticides, nitro-polynuclear aromatic hydrocarbons (PAHs), and hormones (19-nortestosterone, testosterone, 17-methyltestosterone, diethylstilbestrol, and medroxyprogesterone acetate). The authors used a double split generator and achieved low limits of detections. Song et al. [19] developed a nano-HPLC-MS method for the analysis of neurotransmitter histamine in water with 0.1 ng/mL as the limit of detection. The same group [49] reported the analyses of about 29 endocrine disrupting compounds, including PAHs, phenols, and pesticides, by nano-HPLC-MS on a C_{18} column of 75 μm i.d. in marine water. The limit of detection reported by these workers was between 0.4 and 118.7 ng/L. The authors analyzed 20 marine samples and only in two of them two were of the 29 compounds (bisphenol A and 4-*n*-nonylphenol) identified and quantified.

Clicq et al. [50] developed a silicon etched, flat-rectangular nano-channel for the acceleration of coumarin dyes (coumarin C440, CAS No. 26093-31-2, and coumarin C450, CAS No. 26078-21-1) in less than 50 milliseconds. The authors reported 106 plates/meter as theoretical plate of the chip. The mobile phase used was a 40% (v/v) mixture of methanol in purified water. The authors reported that the retention factor was nearly independent of the flow velocity as shown in Fig. 6.7, indicating a variation in retention factor (k) for coumarin 440 (first peak) and 450 (second peak) with the mobile phase velocity. He [51] used isocratic anion-exchange ion chromatography on-chip (8 mm long, 100 μm wide, and 25 μm high) having PS-DVB resins for separation of fluoride, chloride, nitride, bromide, nitrate, phosphate, and sulfate. The mobile phase used was 1.7 mM sodium bicarbonate-1.8 mM sodium carbonate solution with 0.2 μL/min flow rate.

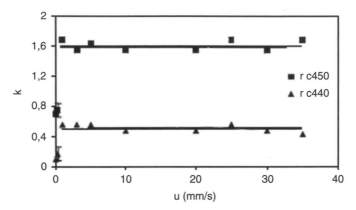

Figure 6.7 Relationship between retention factor (k) and mobile phase velocity (u) for coumarin C440, CAS No. 26093-31-2, and coumarin C450 [50].

All the ions studied were successfully separated and the chromatogram was obtained with the on-chip conductivity detector (Fig. 6.8). The peaks were sharp due to the small column dimensions and minimized dead volume of the integrated system and separation occurred within 90 seconds. Other important applications of nano-HPLC are summarized in Table 6.1.

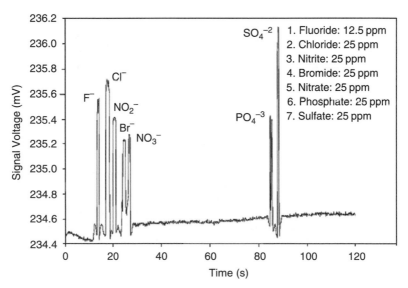

Figure 6.8 Chromatogram of seven anions using on-chip ion chromatography, mobile phase: 1.7 mM sodium bicarbonate and 1.8 mM sodium carbonate with conductivity detection [51].

TABLE 6.1 The Applications of Nano-HPLC for Separation and Identification of Different Molecules

Compounds	Detectors	Chip Materials	LOD	Refs.
Asialoglycans	LED-CFD	PMMA	2 nM	[55]
Actin	LIF	PMMA	30 nM	[56]
Alanine	CCD, LIF, ECD	PMMA, glass, PDMS, polycarbonate	8 nm–7.1 μM	[57–60]
Aspartate	LIF	PDMS, glass	60 nM–200 nM	[60–62]
Anti-estradiol	Scanning-LIF	Glass	4.3 nM	[60]
Alcohol dehydrogenase	LIF	Glass, fused silica	100 fM–0.1 mg/mL	[63–65]
Atropine	CL	Glass	3.8 nM	[66]
Botulinum toxin	LIF	PDMS, poly(IBA), poly(HEMA) agarose, polycarbonate	50 ng	[67]
6-Carboxy-fluorescein	ICCD	Polycarbonate-PMMA	30 nM	[68]
BSA	LIF, MALDI-MS ECD, SPR	PDMS, glass, fused silica, PMMA-CD, polycarbonate	100 fM	[56,58,69–74]
Carbohydrate antigen	ECD	Glass-silicon	2 ng/mL	[75]
Carbonic anhydrase	CCD, LIF	PMMA, glass, fused silica	10 nM	[59,63,64]
Carcinoembryonic antigen	ECD	Glass-silica	2 ng/mL	[75]
Cardiac troponin	Uv-Vis, LIF	Polyacrylimide, PDMS, Silicon	30 ng/mL	[76,77]
Chicken ovalbumin	LIF	Glass	300 pg	[18]
Cy 5	LIF	Glass (Borofloat)	3.3 nM	[78]
Conconavalin A	LIF	PMMA	30 nM	[56]

(Continued)

157

TABLE 6.1 The Applications of Nano-HPLC for Separation and Identification of Different Molecules (*Continued*)

Compounds	Detectors	Chip Materials	LOD	Refs.
C-Reactive protein	Light scattering LIF	PMMA, silicon	1 ng/mL	[77,79,80]
Fluorescein	LIF,UV, ICCD AOD, LIF	PDMS, polycarbonate-PMMA, glass	30 pM	[57,68,81–86]
Dopamine	SVD, ECD, LIF	PDMS, quartz, glass	60 nM	[61,62,87–89]
DsRed	FCS	PMMA	5 nM	[90]
Epinephrine	SVD	PDMS-quartz	500 nM	[87]
Ferritin	ECD	Glass-silica	2 ng/mL	[75]
FLAG peptides	FRET	PDMS	0.48 nM	[91]
IgG (mouse)	LIF	PDMS	15.6 ng/mL	[92]
GABA	LIF	PDMS	60 nM	[61,62]
Glumate	LIF	PDMS, glass	10–200 nM	[62,93]
IgG	GMR, sensors, LIF	BCB, glass	7.0 pM	[94]
Glycine	LIF, ICCD, scanning-LIF	PDMS, polycarbonate-PMMA, glass	60 nM	[59,62,95]
Helix pomatia lectin	LIF	PMMA	30 nM	[56]
Glutamine	CCD, ECD	PMMA, polycarbonate	10 nM	[58,59]
Lysozyme	CCD, deep UV-LIF, LIF	PMMA, quartz, PDMS	10 nM	[59,72,96]
IgG (rabbit)	LIF	PDMS	244 pg/mL	[97]
Immunosuppressive acidic prot.	CL	Quartz	100 nM	[98]
Insulin	LIF, ECD, ICCD, scanning-LIF	Glass, polycarbonate-PMMA	10 pM – 30 nM	[68,99–102]
Interferon-γ	TLS	Glass	1 ng/mL, 60 pM	[103]
Interleukin-6	LIF	Silicon	1 ng/mL	[79]
Isoproterenol	SVD	PDMS-quartz	515 nM	[87]

158

Ovamucoid	LIF	Glass	300 pg	[18]
Lectin peanut agglutin n	LIF	PMMA	30 nM	[56]
Lactalbumin	LIF	Glass	20 nM	[63]
Myoglobin	LIF	PDMS, silicon	30 ng/mL	[77]
Myosin	CCD, LIF	PMMA, glass	100 fM–10 nM	[59,65]
Ovalbumin	CCD, deep UV-LIF, LIF, scanning-LIF	PMMA, quartz, glass	6.4 nM–30 nM	[56,59,81,96]
Serine	LIF, ICCD	PDMS, glass, polycarbonate-PMMA	10 pM–60 nM	[57,61,62,68,104]
Pethidine	CL	Glass	77 nM	[66]
Phosphoproteins	MALDI-MS	PMMA-CD	15 fmol	[74]
Phosphorylase B	CCD	PMMA	10 nM	[59]
Protein A	LIF	PMMA	30 nM	[56]
Staphylococal enterotoxin B	LIF	Glass	28.5 fg/mL	[105]
Serum albumin	CCD	PMMA	10 nM	[59]
Wheat germ aggulatin	LIF	PMMA	30 nM	[56]
Streptavidin	LIF	PMMA	30 nM	[56]
Tacrolimus	LIF	PDMS	5 ng/mL	[106]
Taurine	LIF	PDMS	60 nM	[61,62]
Tetanus neurotoxin	LIF	Glass	2 nM	[107]
Transferrin	LIF	PMMA	30 nM	[56]
Trypsin inhibitor	CCD-LIF, LIF	PMMA, glass	100 fM–20 nM	[59,64,108]
Tumor necrosis factor (TNFα)	LIF	PDMS, silicon	20 pg/mL, 1.14 pM	[109]
Turkey ovalbumin	LIF	Glass	300 pg	[18]

(Continued)

TABLE 6.1 The Applications of Nano-HPLC for Separation and Identification of Different Molecules (*Continued*)

Compounds	Detectors	Chip Materials	LOD	Refs.
Valine	CCD, LIF, ECD	PMMA, glass, polycarbonate	10 nM – 10 μM	[57–59]
β-Galactosidase	CCD, Vis, LIF, LIF	PMMA, acrylic, mylar glass	100 fM – 5 mg/mL	[110, 111]
Xylenecyanol	TLS	Quartz	40 nM	[112]
α-Fetoprotein	ECD	Glass, silicon	2 ng/mL	[75]
Glucose oxidase	SPR	PDMS	11 pM	[69]
β-Human choriogonadotropin	ECD	Glass, silica	2 ng/mL	[75]

Abbreviations: AOD, Acousto-optical deflection; BCB, bisbenzyocyclobutadiene; CCD, indirect contact conductivity detection; CL, chemiluminescence; ECD, electron capture detector; FCS, fluorescence correlation spectroscopy; FRET, fluorescence resonance energy transfer; ICCD, integrated contact conductivity detection; GMR, giant magnetoresistive; LED-CFD, light emitting diode confocal fluorescence detector; LIF, laser-induced fluorescence; LOD, limit of detection; MALDI, matrix-assisted laser desorption ionization; PDMS, poly(dimethylsiloxane); PMMA, poly(methylmethacrylate); SPR, surface plasmon resonance; SVD, sinusoidal voltammetric detection; TLS, thermal lens spectroscopy.

6.4. OPTIMIZATION OF SEPARATIONS IN NANO-HPLC

The optimization of any chromatography modality is the most important aspect in separation science. First of all the selection of the chip-based columns is important followed by its coating materials. Few materials have been used, including reversed phase silicas and glasses but the best separation was achieved on former adsorbent. The composition of the mobile phase, its pH, temperature, amount injected, and detection are the most important factors in getting the best separations. There is no need for special attention in optimization of experimental conditions in nano-HPLC but the methods are similar to those adopted in conventional HPLC. Interested readers should consult our earlier books on this issue [52–54].

6.5. TROUBLESHOOTING IN NANO-HPLC

Nano-HPLC is a very sensitive instrument and needs careful operation especially in sample and mobile phase preparation. The presence of even sub-microscopic solids in a sample may clog the chip column. The flow rate and high pressure on the column often lead to the destruction of the column. Sometimes, overloading may create a problem in separation and require a long time to overcome. A precolumn is always useful to improve the performance of a nano-HPLC system. Not much is known about troubleshooting of chip-based nano-HPLC as the technique is recently developed and more research still underway. These problems may be overcome by proper handling of sample and mobile phase preparation. Besides, a long blank run may get rid of the column saturation. Briefly, nano-HPLC needs a patient scientist for getting effective, efficient, and reproducible results.

6.6. CONCLUSION

Nano-HPLC is a new development in liquid chromatography and is useful in proteomic and genomic research. Besides, it has a good scope in drugs development and design especially in determining in vivo pharmacodynamics and kinetics of drugs. The operation of this system is quite difficult as a minute error may change the results due to nanoscale analyses. All the analytical data of nano-HPLC should be integrated by calculating errors, standard deviation, correlation coefficients, and confidence levels. Briefly, nano-HPLC has a bright future as it will emerge as an efficient analytical method of choice in the present century.

REFERENCES

1. D.R. Reyes, D. Iossifidis, P.A. Auroux, A. Manz, *Anal. Chem.*, **74**, 2623 (2002).
2. T. Vilkner, D. Janasek, A. Manz, *Anal. Chem.*, **76**, 3373 (2004).
3. P.S. Dittrich, K. Tachikawa, A. Manz, *Anal. Chem.*, **78**, 3887 (2006).
4. J.P.C. Vissers, H.A. Claessens, C.A. Cramers, *J. Chromatogr. A*, **1**, 779 (1997).
5. J.P.C. Vissers, *J. Chromatogr. A*, **856**, 117 (1999).
6. J.H. Borges, Z. Aturki, A. Rocco, S. Fanali, *J. Separation Sci.*, **30**, 1589 (2007).
7. S. Koster, E. Verpoorte, *Lab. Chip*, **7**, 1394 (2007).
8. G.T. Roman, R.T. Kennedy, *J. Chromatogr. A*, **1168**, 170 (2007).
9. Y. Ishihama, *J. Chromatogr. A*, **1067**, 73 (2005).
10. F. Svec, C.G. Huber, *Anal. Chem.*, **78**, 2100 (2006).
11. F. Qin, C. Xie, Z.Yu, L. Kong, *J. Separation Sci.*, **29**, 1332 (2006).
12. M. McEnery, J.D. Glennon, J. Alderman, S.C. O'Mathuna, *J. Capil. Electrophor. Microchip Technol.*, **6**, 33 (1999).
13. C. Ericson, J. Holm, T. Ericson, S. Hjerten, *Anal. Chem.* **72**, 81 (2000).
14. B. He, N. Tait, F. Regnier, *Anal. Chem.*, **70**, 3790 (1998).
15. G. Ocvirk, E. Verpoorte, A. Manz, M. Grasserbauer, H.M. Widmer, *Anal. Methods Instrum.*, **2**, 74 (1995).
16. L. Licklider, X.Q. Wang, A. Desai, Y.C. Tai, T.D. Lee, *Anal. Chem.*, **72**, 367 (2000).
17. A.R. Ivanov, L. Zang, B.L. Karger, *Anal. Chem.*, **75**, 5306 (2003).
18. X. Mao, Y. Luo, Z. Dai, K. Wang, Y. Du, B. Lin, X. Mao, Y. Luo, Z. Dai, K. Wang, Y. Du, B. Lin, *Anal. Chem.*, **76**, 6941 (2004).
19. Y. Song, Z. Quan, Y.M. Liu, *Rapid Commun. Mass Spectrom.*, **18**, 2818 (2004).
20. M. Bedair, Z. El Rassi, *J. Chromatogr. A*, **1044**, 177 (2004).
21. F.M. Okanda, Z. El Rassi, *Electrophoresis*, **27**, 1020 (2006).
22. H. Yin, K. Killen, R. Brennen, D. Sobek, M. Werlich, T. Van de Goor, *Anal. Chem.*, **77**, 527 (2005).
23. F. Vanroabaeys, R. Van-Coster, G. Dhondt, B. Devreese, J. Van Beeumen, *J. Prot. Res.*, **4**, 2283 (2005).
24. V. Tschäppät, E. Varesio, L. Signor, G. Hopfgartner, *J. Separation Sci.*, **28**, 1704 (2005).
25. M. Ghitun, E. Bonneil, M.H. Fortier, H. Yin, K. Killeen, P. Thibault, *J. Separation Sci.*, **29**, 1539 (2006).
26. J.T.J. Huang, T. McKenna, C. Hughes, F.M. Leweke, E. Schwarz, S. Bahn, *J. Separation Sci.*, **30**, 214 (2007).
27. G.E. Yue, MG. Roper, C. Balchunas, A. Pulsipher, J.J. Coon, J. Shabanowitz, D.F. Hunt, J.P. Landers, J.P. Ferrance, *Anal. Chim. Acta*, **564**, 116 (2006).

28. Q. Wang, L. Li, D.R. Li, W. Zhang, X. Wei, J.Q. Zhang, Y. Tang, *Zhonghua Fu Chan Ke Za Zhi.*, **41**, 544 (2006).

29. A. Ishida, T. Yoshikawa, M. Natsume, T. Kamidate, *J. Chromatogr. A.*, **1132**, 90 (2006).

30. J. Srbek, J. Eickhoff, U. Effelsberg, K. Kraiczek, T. van de Goor, P. Coufal, *J. Separation Sci.*, **30**, 2046 (2007).

31. M. Gaspari, V. Abbonante, G. Cuda, *J. Separation Sci.*, **30**, 2210 (2007).

32. K. Lanckmans, B. Stragier, S. Sarre, I. Smolders, Y. Michotte, *J. Separation Sci.*, **30**, 2217 (2007).

33. M. Vollmer, P. Hörth, G. Rozing, Y. Couté, R. Grimm, D. Hochstrasser, J.C. Sanchez, *J. Separation Sci.*, **29**, 499 (2006).

34. A. Staes, E. Timmerman, J. Van Damme, K. Helsens, J. Vandekerckhove, M. Vollmer, K. Gevaert, *J. Separation Sci.*, **30**, 1468 (2007).

35. J. Hardouin, M. Duchateau, R. Joubert-Caron, M. Caron, *Rapid Commun. Mass Spectrom*, **20**, 3236 (2006).

36. J. Hardouin, R. Joubert-Caron, M. Caron, *J. Separation Sci.*, **30**, 1482 (2007).

37. J. Li, T. LeRiche, T.L. Remblay, C. Wang, E. Bonneil, D.J. Harrison, P. Thibault, *Mol. Cell Proteomics*, **1**, 157 (2002).

38. M.S. Lee, E.H. Kerns, *Mass Spectrom. Rev.*, **18**, 187 (1999).

39. D. Wistuba, V. Schurig, *J. Chromatogr. A*, **875**, 255 (2000).

40. K. Lanckmans, A. Van Eeckhaut, S. Sarre, I. Smolders, Y. Michotte, *J. Chromatogr. A*, **1131**, 166 (2006).

41. E. Chmela, R. Tijssen, M.T. Blom, H.J. Gardeniers, A. van den Berg, *Anal. Chem.*, **74**, 3470 (2002).

42. M.T. Blom, E. Chmela, R.E. Oosterbroek, R. Tijssen, A. van den Berg, *Anal Chem.*, **75**, 6761 (2003).

43. C.Y. Shih, Y. Chen, J. Xie, Q. He, Y.C. Tai, *J. Chromatogr. A.*, **1111**, 272 (2006).

44. S. Fanali, Z. Aturki, G. D'Orazio, A. Rocco, *J. Chromatogr. A*, **1150**, 252 (2007).

45. H.V. Fuentes, A.T. Woolley, *Lab. Chip*, **7**, 1524 (2007).

46. C.Y. Lu, C.H. Feng, *J. Liq. Chromatogr. Rel. Technol.*, **31**, 54 (2008).

47. A. Cappiello, G. Famiglini, P. Palma, F. Mangani, *Anal. Chem.*, **74**, 3547 (2002).

48. A. Cappiello, G. Famiglini, F. Mangani, P. Palma, A. Siverio, *Anal. Chim. Acta*, **493**, 125 (2003).

49. G. Famiglini, P. Palma, A. Siviero, M.A. Rezai, A. Cappiello, *Anal. Chem.*, **77**, 7654 (2005).

50. D. Clicq, S. Vankrunkelsven, W. Ranson, C. De Tandt, G.V. Baron, G. Desmet, *Anal. Chim. Acta*, **507**, 79 (2004).

51. Q. He, *Integrated nano liquid chromatography system on a chip*, Ph.D. Thesis, California Institute of Technology, Pasadena, California, June 21 (2005).

52. I. Ali, H.Y. Aboul-Enein, *Instrumental methods in metal ions speciation: Chromatography, capillary electrophoresis and electrochemistry*, New York: Taylor & Francis (2006).

53. I. Ali, H.Y. Aboul-Enein, *Chiral pollutants: Distribution, toxicity and analysis by chromatography and capillary electrophoresis*, Chichester: Wiley (2004).

54. H.Y. Aboul-Enein, I. Ali, *Chiral separations by liquid chromatography and related technologies*, New York: Marcel Dekker (2003).

55. F. Dang, K. Kakehi, J. Cheng, O. Tabata, M. Kurokawa, K. Nakajima, M. Ishikawa, Y. Baba, *Anal. Chem.*, **78**, 1452 (2006).

56. H. Shadpour, S.A. Soper, *Anal. Chem.*, **78**, 3519 (2006).

57. C.T. Culbertson, S.C. Jacobson, J.D. Ramsey, *Anal. Chem.*, **72**, 5814 (2000).

58. H. Shadpour, M.L. Hupert, D. Patterson, C. Liu, M. Galloway, W. Stryjewski, J. Goettert, S.A. Soper, *Anal. Chem.*, **79**, 870 (2007).

59. M. Galloway, W. Styjewski, A. Henry, S.M. Ford, S. Llopis, R.L. McCarley, S.A. Soper, *Anal. Chem.*, **74**, 2407 (2002).

60. G.T. Roman, T. Hlaus, K.J. Bass, T.G. Seelhammer, C.T. Culbertson, *Anal. Chem.*, **77**, 1414 (2005).

61. N.A. Cellar, S.T. Burns, J.C. Meiners, H. Chen, R.T. Kennedy, *Anal. Chem.*, **77**, 7067 (2005).

62. N.A. Cellar, R.T. Kennedy, *Lab. Chip*, **6**, 1205 (2006).

63. A.E. Herr, A.K. Singh, *Anal. Chem.*, **76**, 4727 (2004).

64. T.T. Razunguzwa, M. Warrier, A.T. Timperman, *Anal. Chem.*, **78**, 4326 (2006).

65. R.S. Foote, J. Khandurina, S.C. Jacobson, J.M. Ramsey, *Anal. Chem.*, **77**, 57 (2005).

66. P.A. Greenwood, M. Carolyn, T. McCreedy, G.M. Greenway, *Talanta*, **56**, 539 (2002).

67. J. Moorthy, G.A. Mensing, D. Kim, S. Mohanty, D.T. Eddington, W.H. Tepp, E.A. Johnson, D.J. Beebe, *Electrophoresis*, **25**, 1705 (2004).

68. J.G. Shackman, M.S. Munson, D. Ross, *Anal. Chem.*, **79**, 565 (2007).

69. N. Ly, K. Foley, N. Tao, *Anal. Chem.*, **79**, 2546 (2007).

70. B.E. Slentz, N.A. Penner, F.E. Regnier, *J. Chromatogr. A*, **984**, 97 (2003).

71. S.M. Kim, M.A. Burns, E.F. Hasselbrink, *Anal. Chem.*, **78**, 4779 (2006).

72. D. Xiao, T. Van Le, M.J. Wirth, *Anal. Chem.*, **76**, 2055 (2004).

73. H. Nagata, M. Tabuchi, K. Hirano, Y. Baba, *Electrophoresis*, **26**, 2687 (2005).

74. M. Gustafsson, D. Hirschberg, C. Palmberg, H. Jornvall, T. Bergman, *Anal. Chem.*, **76**, 345 (2004).

75. M.S. Wilson, W. Nie, *Anal. Chem.*, **78**, 6476 (2006).

76. J.H. Cho, S.M. Han, E.H. Paek, I.H. Cho, S.H. Paek, *Anal. Chem.*, **78**, 793 (2006).

77. M. Wolf, D. Juncker, B. Michel, P. Hunziker, E. Delamarche, *Biosens. Bioelectron.*, **19**, 1193 (2004).

78. J.C. Toulet, R. Volkel, H.P. Herzip, E. Verpoorte, N.F.D. Rooij, R. Dandliker, *Anal. Chem.*, **74**, 3400 (2002).

79. N. Christodoulides, M. Tran, P.N. Floriano, M. Rodriguez, A. Goodey, M. Ali, D. Neikirk, J.T. McDevitt, *Anal. Chem.*, **74**, 3030 (2002).

80. N. Pamme, R. Koyama, A. Manz, *Lab. Chip*, **3**, 187 (2003).

81. S.B. Cheng, C.D. Skinner, J. Taylor, S. Attiya, W.E. Lee, G. Picelli, D.J. Harrison, *Anal. Chem.*, **73**, 1472 (2001).

82. M. Bowden, L. Song, D.R. Walt, *Anal. Chem.*, **77**, 5583 (2005).

83. J.P. Shelby, D.T. Chiu, *Anal. Chem.*, **75**, 1387 (2003).

84. G. Ocvirk, M. Munroe, T. Tang, R. Oleschuk, K. Westra, D.J. Harrison, *Electrophoresis*, **21**, 107 (2000).

85. M.W. Li, B.H. Huynh, M.K. Hulvey, S.M. Lunte, R.S. Martin, *Anal. Chem.*, **78**, 1042 (2006).

86. J.C. Sanders, Z. Huang, J.P. Landers, *Lab. Chip*, **1**, 167 (2001).

87. N.E. Hebert, B. Snyder, R.L. McCreery, W.G. Kuhr, S.A. Brazill, *Anal. Chem.*, **75**, 4265 (2003).

88. C.C. Wu, R.G. Wu, J.G. Huang, Y.C. Lin, H.C. Chang, *Anal. Chem.*, **75**, 947 (2003).

89. N.A. Lacher, S.M. Lunte, R.S. Martin, *Anal. Chem.*, **76**, 2482 (2004).

90. P.S. Dittrich, P. Schwille, *Anal. Chem.*, **74**, 4472 (2002).

91. M.E. Piyasena, T. Buranda, Y. Wu, J. Huang, L.A. Sklar, G.P. Lopez, *Anal. Chem.*, **76**, 6266 (2004).

92. R.L. Millen, T. Kawaguchi, M.C. Granger, M.D. Porter, M. Tondra, *Anal. Chem.*, **77**, 6581 (2005).

93. Z.D. Sandlin, M. Shou, J.G. Shackman, R.T. Kennedy, *Anal. Chem.*, **77**, 7702 (2005).

94. R.L. Millen, T. Kawaguchi, M.C. Granger, M.D. Porter, M. Tondra, *Anal. Chem.*, **77**, 6581 (2005).

95. H. Xu, T.P Roddy, J.A. Lapos, A.G. Ewing, *Anal. Chem.*, **74**, 5517 (2002).

96. P. Schultze, M. Ludwig, F. Kohler, D. Belder, *Anal. Chem.*, **77**, 1325 (2005).

97. K.S. Kim, J.K. Park, *Lab. Chip*, **5**, 657 (2005).

98. K. Tsukagoshi, N. Jinno, R. Nakajima, *Anal. Chem.*, **77**, 1684 (2005).

99. J.F. Dishinger, R.T. Kennedy, *Anal. Chem.*, **79**, 947 (2007).

100. J.G. Shackman, G.M. Dahlgren, J.L. Peters, R.T. Kennedy, *Lab. Chip*, **5**, 56 (2004).

101. J. Wang, A. Ibanez, M.P. Chatrathi, *J. Am. Chem. Soc.*, **125**, 8444 (2003).

102. M.G. Roper, J.G. Shackman, G.M. Dahlgren, R.T. Kennedy, *Anal. Chem.*, **75**, 4711 (2003).

103. K. Sato, M. Yamanaka, T. Hagino, M. Tokeshi, H. Kimura, T. Kitamori, *Lab. Chip*, **4**, 570 (2004).

104. G.T. Roman, K. McDaniel, C.T. Culbertson, *Analyst*, **131**, 194 (2006).

105. A.J. Haes, A. Terray, G.E. Collins, *Anal. Chem.*, **78**, 8412 (2006).

106. Y. Murakami, T. Endo, S. Yamamura, N. Nagatani, Y. Takamura, E. Tamiya, *Anal. Biochem.*, **334**, 111 (2004).

107. A.E. Herr, D.J. Throckmorton, A.A. Davenport, A.K. Singh, *Anal. Chem.*, **77**, 585 (2005).

108. A.E. Herr, A.K. Singh, *Anal. Chem.*, **76**, 4727 (2004).

109. S. Cesaro-Tadic, G. Dernick, D. Juncker, G. Burrman, H. Kropshofer, B. Michel, C. Fattinger, E. Delamarche, *Lab. Chip*, **4**, 563 (2004).

110. T.T. Razunguzwa, M. Warrier, A.T. Timperman, *Anal. Chem.*, **78**, 4326 (2006).

111. Y. Mizukami, D. Rajniak, A. Rajniak, M. Nishimura, *Sensors & Actuators B Chem.*, **81**, 202 (2002).

112. E.A. Schilling, A.E. Kamholz, P. Yager, *Anal. Chem.*, **74**, 1798 (2002).

113. K. Mawatari, Y. Naganuma, K. Shimoide, *Anal. Chem.*, **77**, 687 (2005).

CHAPTER 7

NANOCAPILLARY ELECTROCHROMATOGRAPHY AND NANOMICELLAR ELECTROKINETIC CHROMATOGRAPHY

7.1. INTRODUCTION

In addition to HPLC, microchips have also been used in other modalities of liquid chromatography, including capillary electrochromatography and micellar electrokinetic chromatography. Many workers have attempted to achieve nano separations at high speed of different molecules with high efficiency, reproducibility, and low detection limits. The state of the art of separation in these modalities is discussed in this chapter, with special emphasis on their applications, optimization, and mechanisms of separation.

7.2. NANOCAPILLARY ELECTROCHROMATOGRAPHY

Basically, capillary electrochromatography (CEC) is a hybrid technique of HPLC and CE [1–3], which was developed in 1974 by Pretorius et al. [4]. CEC is expected to combine high peak efficiency, which is a characteristic of electrically driven separations, with high separation selectivity. As is the case for electrophoresis, a voltage is applied across the separation plateform and sample moves via electroosmotic flow (EOF). However, in analogy to liquid chromatography, the separation device contains a solid

Nanochromatography and Nanocapillary Electrophoresis. By Ali, Aboul-Enein, and Gupta
Copyright © 2009 John Wiley & Sons, Inc.

stationary phase providing the medium for solute interactions. Therefore, electrochromatographic separations are achieved as a result of differences in both electrophoretic mobility and specific interactions. Normal CEC experiments are carried out on wall coated open tubular capillaries or capillaries packed with particulate or monolithic silica or other inorganic materials as well as organic polymers. CEC uses a long capillary as the separation platform, which, recently, has been replaced by a small capillary on a microchip. This works with all experimental conditions at nanolevels, that is, amount injected, flow rate, detection, etc. Therefore, in this book we term CEC, with a capillary on a microchip, nanocapillary electrochromatography (NCEC).

NCEC is emerging as an attractive alternative to nanocapillary electrophoresis (NCE) and nanoliquid chromatography (NLC) due to the relative ease with which it can be miniaturized into a microfluidic chip format. In NCE, separation is achieved by applying an electric field and, since fluid flow is driven by EOF, the separation device does not need pumps or valves. Besides, the flat flow profile generated by EOF minimizes dispersion of analyte bands during their passage through the stationary phase, allowing very high plate counts to be achieved. Since EOF is largely independent of channel or particle size, both small size beads or monolithic stationary phases with very small pores can be used, facilitating solute mass transfer without generating large pressure drops as is the case for NLC. In 1994 Jacobson et al. [5] published the first paper on NCEC and since then it has gained much importance in the nano world of analytical science. As discussed above, NCEC has been used for the separation of a variety of compounds and some reviews [6–19] have also been published in this area. In these reviews, NCEC has been discussed in terms of fabrication of the channel, coating of stationary phases, advantages and disadvantages, optimization, detector hyphenation, and applications. The state of the art of NCEC separations is discussed in the following sections.

7.2.1 Biological Samples

NCEC is supposed to be the best analytical technique in proteomic, genomic, and drug development programs due to its inherent character of nano nature. Normally, amounts of samples in these disciplines are in nanoliters, which need this sort of device. Therefore, this modality of liquid chromatography has been used for the separation and identification of various molecules in different biological matrices.

Ceriotti and Verpoorte [20] integrated a fritless column for NCEC with conventional stationary phases, which was used for the separation of fluorescein isothiocyanate (FITC)-labeled amino acids. The chips were fabricated in poly(dimethylsiloxane) using deep-reactive-ion-etched silicon masters. The

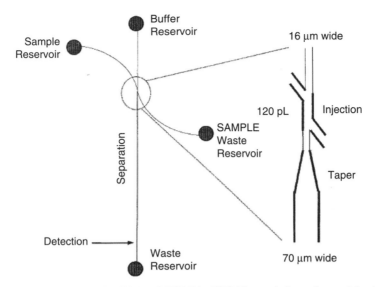

Figure 7.1   Chip layout for fritless NCEC in PDMS consisting of two side channels entering into the main channel to form slanted T intersections [20].

microchannel was coated with 3 μm octadecylsilanized silica microspheres, which were stabilized by a thermal treatment. The stability and quality were evaluated using in-column indirect fluorescence detection. The effects of voltage on EOF and on efficiency were investigated. The layout of the structures used by these authors is shown in Fig. 7.1, which represents an injection element consisting of two 16 μm wide channels entering into the main straight channel to form two offset, slanted T junctions. The separation channel

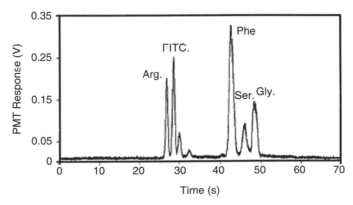

Figure 7.2   Electropherogram of a mixture of FITC-labeled amino acids, achieved in an oxidized PDMS/glass device [20].

included a gradual tapering at the top of the separation column just below the intersection. The authors used the CleWin (Delta Mask, Enschede, The Netherlands) program for the design of the layout. An electropherogram of a mixture of $10 \, \mu m$ FITC-labeled amino acids, achieved on an oxidized PDMS/glass device, is shown in Fig. 7.2, with a 100 mM Tris-20 mM boric acid buffer (pH 9). Furthermore, the authors attempted to optimize the separation. Figure 7.3 indicates a linear relationship between field strength and linear velocity.

Park et al. [21] developed nanocapillary electrochromatographic separations of FITC derivatized amino acids. Because of a large surface area to volume ratio of the silica packing, reproducible control of EOF was observed without leveling of the solutions in the reservoirs. The chromatogram of the FITC-derivatized amino acid mixture is shown in Fig. 7.4 on 2.5 mm channel cross with 20 mM $Na_2B_4O_7$ running buffer (pH 9.2), and 200 V/cm as applied voltage. Furthermore, these authors studied the effect of channel length sampling and the results obtained are shown in Fig. 7.5. It may be seen from this figure that the separation was poor at 2 mm and severe band broadening occurred at 3 mm. As a result 2 mm from the channel cross was the best sampling condition.

Faure et al. [22] described nanoelectrochromatography on poly(dimethyl)-siloxane microchips using organic monolithic stationary phases for analysis of derivatized catecholamines. Surface modification of the PDMS material was carried out by UV-mediated graft polymerization. The efficiency of the unit was ascertained by measuring theoretical plates, which were 200,000 per meter. Furthermore, the authors optimized the separation by using pinched and electrokinetic modes at different applied potential of 1.0 kV/cm (Fig. 7.6) and 30 kV/cm (Fig. 7.7) for the pinched and electrokinetic

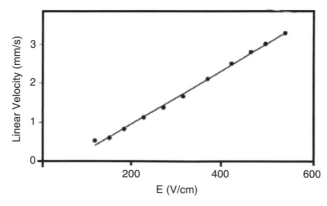

Figure 7.3 Plot of mobile phase linear velocity versus applied electric field strength [20].

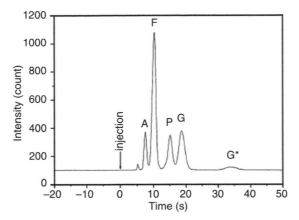

Figure 7.4 Chromatogram of FITC-derivatized amino acid mixture. A, arginine; F, FITC; P, phenylalanine; G, glycine; G*, glutamic acid [21].

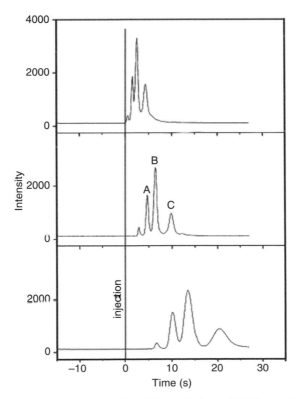

Figure 7.5 Effect of detection position (A) 1, (B) 2, and (C) 3 mm from the channel cross [21].

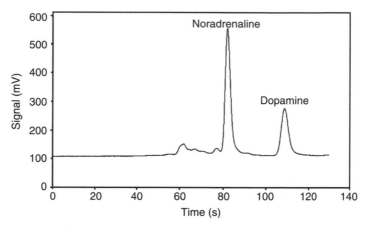

Figure 7.6 Electrochromatogram of noradrenaline and dopamine, respectively. Pinched injection for 50 seconds (S 1 kV; SW ground; B and BW 0.6 kV) [22].

modes, respectively. The buffer used was 40% sodium phosphate (5 mM, pH 2.0) and 60% acetonitrile. A perusal of these figures indicates the best separation in Fig. 7.7.

Slentz et al. [23] described nanocapillary electrochromatography on collocated monolithic support structures (COMOSS) molded in PDMS. The authors fabricated a chromatographic channel by molding COMOSS directly. In addition, the ability to separate biological samples such as peptides from a

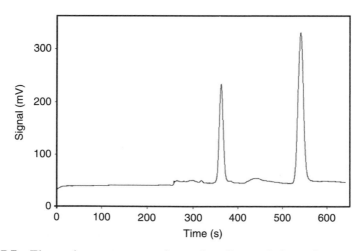

Figure 7.7 Electrochromatograms of noradrenaline and dopamine, respectively. Electrokinetic injection at 20 kV/cm for 2 seconds followed by separation at 30 kV/cm, L = 54 cm (total length of capillary), l = 40 cm (length of capillary up to detector) [22].

tryptic digest of FITC-labeled bovine serum albumin (FITC-BSA) was evaluated. The separation efficiency was measured in terms of theoretical plates, which were 4.0×10^5 plates/meter. Furthermore, the same group [24] described the separation of FITC-labeled peptides (FITC–Gly–Phe–Glu–Lys(FITC)–OH, FITC–Gly–Phe–Glu–Lys–OH, FITC and FITC–Gly–Tyr–OH) on a C_{18}-AMPS modified PDMS COMOSS microchip. Electrokinetic injection at 1000 V/cm for 0.25 s were applied with mobile phase 1.0 mM carbonate buffer (pH 9.0). The applied voltage was 3.0 kV/cm with fluorescence detection. The separation of these peptides was carried out on various microchips grafted with various stationary phases: poly(vinylsulfonic acid) (VSA), polyacrylic acid (PAA), and poly(styrenesulfonic acid (PS-SA) and the best separation was achieved on a PS-SA coated chip (Fig. 7.8). The same group [25] optimized the separations in NCEC by optimizing voltage for the separation of rhodamine B on COMOSS chip using 1.0 mM sodium hydrogen carbonate-potassium carbonate buffer (pH 9.0) as mobile phase.

Gottschlich et al. [26] described the separation of tetramethyl rhodamine isothiocyanate (TRITC)-labeled tryptic peptides of β-casein. The field strength was 220 V/cm in the NCEC channel with 10 mM sodium borate with 30% (v/v) acetonitrile as mobile phase. Throckmorton et al. [27] described the separation of papain inhibitor, proctolin, opioid peptide (α-casein fragment 90–95), Ile-angiotensin III and angiotensin III on a porous polymer monolith

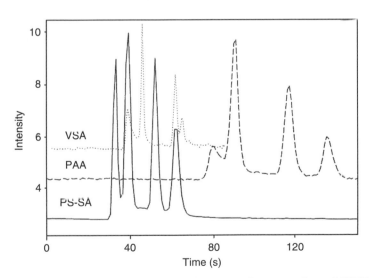

Figure 7.8 Effect of various stationary phases on the separation of FITC-labeled peptides. VSA, poly(vinylsulfonic acid); PAA, polyacrylic acid; and PS-SA, poly(styrenesulfonic acid) [24].

filled microchip. The polymer was negatively charged lauryl acrylate mono-lith, peptides were labeled with DNA, and fluorescence detection was used. The background electrolyte (BGE) applied was acetonitrile-25 mM borate (pH 8.2) containing 10 mM octanesulfonate (30:70) with 5.0 kV/cm as applied voltage. The chromatogram is shown in Fig. 7.9 indicating good resolution of these molecules.

Lazar et al. [8] developed a microfabricated device for exploiting capillary electrochromatography-mass spectrometry for testing analysis of protein digests and the detection limit achieved was at fmol level. The injector oper-ation was optimized for performing nonuniform EOF within the microfluidic channels. The dimensions of one processing line were sufficiently small to enable the integration of four- to eight-channel multiplexed structures on a single substrate. Jindal and Cramer [28] described nanoelectrochromatography using a sol gel immobilized stationary phase with UV absorbance detection for separation of three peptides (Trp-Ala, Leu-Trp, and Trp-Trp) under isocratic chromatographic conditions. Localization of the stationary phase was achieved by immobilizing the stationary phase in the separation channel prior to bond-ing of the cover plate. An optical fiber set up was developed for carrying out on-chip UV absorbance detection. The effect of applied voltage on velocity was evaluated using thiourea as an unretained marker. Furthermore, the

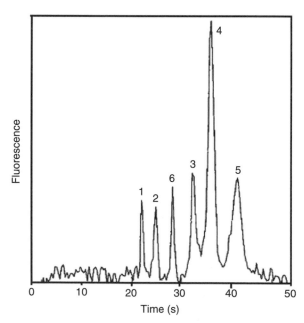

Figure 7.9 Chromatogram of (1) papain inhibitor, (2) proctolin, (3) opioid peptide (α-casein fragment 90-95), (4) Ile-angiotensin III, (5) angiotensin III, and (6) GGG [27].

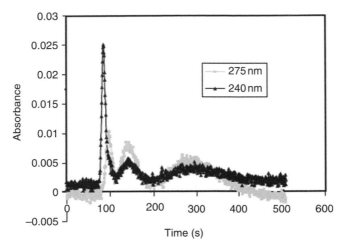

Figure 7.10 Electrochromatogram of a mixture of thiourea and the peptides Trp-Ala, Leu-Trp, and Trp-Trp using C_4 modified silica particles (5 μm, 300 Å pore size) immobilized in the sol gel [28].

effect of immobilization was also observed on the separation of these molecules. This is shown in Fig. 7.10 and Fig. 7.11 on C_4 modified silica particles (5 μm, 300 Å pore size) immobilized in the sol-gel. These figures clearly indicate that the best separation was achieved without any stationary in channel. Besides, efforts were also made by these workers to optimize separations by using 275 and 240 nm as wavelengths (Fig. 7.10 and Fig. 7.11) indicating a low level of detection at 275 nm.

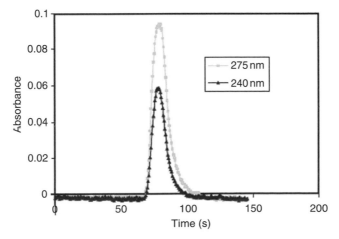

Figure 7.11 Electrochromatogram of a mixture of thiourea and the peptides Trp-Ala, Leu-Trp, and Trp-Trp without stationary phase [28].

Galloway and Soper [29] used a poly-(methyl methacrylate) (PMMA) chip in NCEC for carrying out the separation of two DNA fragments produced via the polymerase chain reaction (PCR) of λ-DNA. The DNA sizing ladder, which contained six DNA fragments ranging in size from 100 to 2000 base pairs, was separated nearly at the baseline, with an efficiency of the order of 10^4 plates for a 3 cm long column. Galloway et al. [30] used open channel NCEC for the separation of a double stranded DNA ladder in plain and PMMA-modified chips (C_{18} silica gel). The mobile phase used was 25% acetonitrile and 75% aqueous phase containing 50 mM TEAA (ion pairing agent, pH 7.4). The applied voltage was 100 V/cm with conductivity detection. Kutter et al. [31] described NCEC for the separation of coumarin 440, coumarin 450, coumarin 460, and coumarin 480 using 10 mM borate buffer (pH 8.4) with a linear gradient of acetonitrile from 29% to 50% within 5 s, starting 1 s after injection. The NCEC separation of these coumarin derivatives is shown in Fig. 7.12, indicating a baseline resolution of all species. Furthermore, the authors studied the effect of amount of acetonitrile on the separation of coumarin (Fig. 7.13) and it was observed that the best resolution was with 29% acetonitrile.

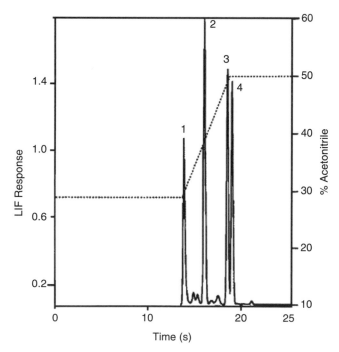

Figure 7.12 Chromatogram of the separation of coumarin derivatives [31].

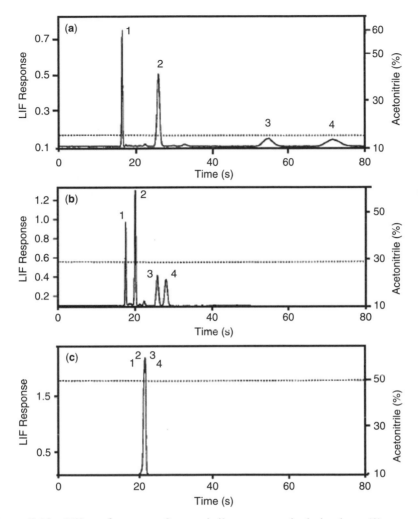

Figure 7.13 Effect of amount of acetonitrile on coumarin derivatives, (1) coumarin 440, (2) coumarin 450, (3) coumarin 460, and (4) coumarin 480 on 5.2 μm; stationary phase, octadecylsilane with mobile phase, 10 mM borate buffer (pH 8.4) with different amounts of acetonitrile, (a) 15%, (b) 29%, and (c) 50% [31].

7.2.2 Environmental Samples

As in the case of biological samples, most pollutants are present at trace levels in our environment. The analyses of such pollutants require nanoseparation techniques and NCEC is a good choice in this area. Some workers have attempted to analyze a few pollutants using this modality of liquid chromatography, which are summarized in the following paragraphs.

Fintschenko et al. [32] described nanoelectrochromatography of polycyclic aromatic hydrocarbons on an acrylate-based UV-initiated porous polymer monolith. The authors discussed the photodefinability of the monolith cast in the channels during the polymerization process. Furthermore, the chromatographic performance of this device was compared with the chips completely filled with monolith. Thirteen polycyclic aromatic hydrocarbons (PAHs) were separated and detected by laser-induced fluorescence detection at 257 nm. It has also been reported that minimum plate height was 5 μm for early-, middle-, and late-eluting compounds, with 200,000 as the average number of theoretical plates per meter. A within-chip variability in the retention time of 2% to 10% relative standard deviation (RSD) was reported and the results demonstrated the feasibility and reliability of the device. Ericson et al. [33] described fast NCE on C_3 porous polymer monolith chip for the separation of uracil, phenol, and benzyl alcohol. The BGE used was 5 mM sodium phosphate (pH 7.4) containing 15% (v/v) acetonitrile with 2.4 kV/cm as applied voltage. The injection mode was electrokinetic at 100 V for 2 seconds. The chromatograms of these compounds are shown in Fig. 7.14 indicating good resolution.

Pumera et al. [34] used gold nanoparticle enhanced open channel NCEC for the separation of p-aminophenol, o-aminophenol, and m-aminophenol in a bare glass channel and polydiallyl dimethyl ammonium chloride (PDADMAC) gold coated channel. The mobile phase used was acetate buffer (20 mM, pH 5.0) with electrokinetic sample injection at 1.5 kV/cm for 3 seconds and 2.0 kV/cm as separation voltage (Fig. 7.15). Broyles

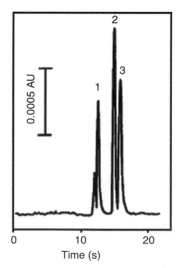

Figure 7.14 Chromatogram of (1) uracil, (2) phenol, and (3) benzyl alcohol on C_3 porous polymer monolith coated chip [33].

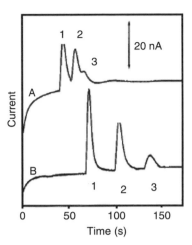

Figure 7.15 Chromatogram of the separation of (1) *p*-aminophenol, (2) *o*-aminophenol, and (3) *m*-aminophenol, (A) without treatment and (B) after PDADMAC gold coating chips [34].

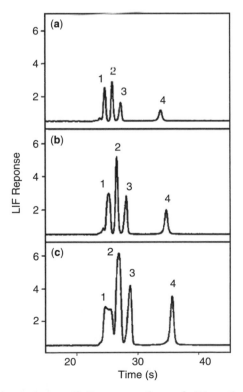

Figure 7.16 Chromatograms of the separation of (1) anthracene, (2) pyrene, (3) 1,2-benzofluorene, and (4) benzo[a]pyrene in NCEC coated with octadecylsilane [35].

et al. [35] reported the separation of anthracene, pyrene, 1,2-benzofluorene, and benzo[a]pyrene in NCEC coated with octadecylsilane. The running buffer was 10 mM Tris at 52% (v/v) acetonitrile, holding for 10 seconds and then switching to 56% (v/v) acetonitrile (Fig. 7.16). The applied voltage on channel was 500 V/cm. Furthermore, the authors optimized separation by using a mobile phase program, that is, isocratic step gradient and linear gradient (Fig. 7.17). The best separation achieved was in step gradient mode. Wang et al. [36] developed reversed phase NCEC by using octadecylsilane

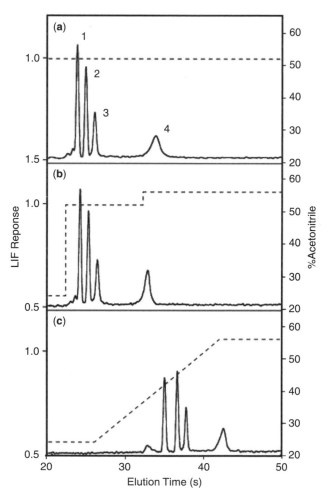

Figure 7.17 Effect of mobile phase program (a) isocratic, (b) step gradient, and (c) linear gradient on the separation of (1) anthracene, (2) pyrene, (3) 1,2-benzofluorene, and (4) benzo[a]pyrene [35].

bonded silica particles for the separation of a mixture of thiourea, toluene, naphthalene, fluorene, and anthracene. NCEC conditions were 50 cm × 250 μm i.d. (5 cm section packed with 3.0 μm ODS magnetic particles), with mobile phase of 5 mM phosphate buffer (pH 8.0) containing 70% methanol. The authors advocated that the magnetic attraction approach to fritless column packing may be used for construction of advanced chip-based NCEC in complex architectures comprising curved and intersecting channels.

7.2.3 Mechanism of Separation

NCEC works on the separation principle of conventional CEC. The chromatographic and electrophoretic mechanisms work simultaneously in NCEC and several combinations are possible. The separation occurs on the mobile phase/stationary phase interface and the exchange kinetics between the mobile and stationary phases are important. Basically, NCEC uses an electroendosmotically driven flow instead of a pressure-driven flow, to propel the mobile phase through the column. Electroendosmotic flow is generated in the electrical double layer at charged solid liquid interfaces. The separation mechanism in NCEC is based primarily on differential interactions such as partition between two phases, adsorption, van der Waal forces, steric hindrance, etc. The charged analytes are also influenced by the electric field, resulting in their differential migration. NCEC offers the same stationary phases with different chromatographic properties and broad application range of retention mechanisms and selectivities. The chromatographic band broadening mechanisms are quite different in individual mode. Sometimes low conductivity buffers can also be employed to suppress zone broadening [37].

7.3. NANOMICELLAR ELECTROKINETIC CHROMATOGRAPHY

Micellar electrokinetic chromatography (MEKC) is a modality of liquid chromatography having a surfactant molecule in the form of a micelle, which was introduced by Terabe et al. in 1984 [38]. The formation and separation occur in the capillary and, hence, it is also called micellar electrokinetic capillary chromatography (MECC). This modality is useful for some specific molecules having solubilities in micelles and, therefore, utilized for the separation and identification of such compounds with great efficiency, reproducibility, and low levels of detections. The most commonly used compounds for micelle formation are sodium dodecyl sulfate (SDS), sodium tetradecyl sulfate, sodium decanesulfonate, sodium N-lauryl-N-methyllaurate, sodium polyoxyethylene dodecyl ether sulfate, sodium N-dodecanoyl-L-valinate, sodium cholate, sodium deoxycholate, sodium taurocholate, sodium taurodeoxycholate, potassium perfluoroheptanoate, dodecyltrimethylammonium chloride, dodecyltrimethylammonium bromide, tetradecyltrimethylammonium bromide, and cetyltrimethylammonium bromide. As in the case of CEC, MEKC is also carried out in long capillaries, which has been replaced by short capillaries on microchips. Again we call this sort of assembly nanomicellar electrokinetic chromatography (NMEKC). Few reviews [39–41] have been cited in the literature highlighting various aspects of NMEKC. Some workers developed NMEKC methods for the separation and identification of different compounds, which are discussed in the following sections along with strategies for optimization.

7.3.1 Biological Samples

NMEKC has been used for analyses of some compounds in biological samples, the most common of which are drugs and proteins. Some examples are given below.

von Heeren et al. [42] described nanomicellar electrokinetic chromatography on a planar glass microchip with laser-induced fluorescence detection for analysis of six FITC-labeled amino acids (Fig. 7.18). The mobile phase used was 10 mM Na_2HPO_4, and 6 mM $Na_2B_4O_7$ (pH 9.2) containing 75 mM SDS, with 3.3 nM as the detection limit. This method was also applied in human urine and serum samples on an uncoated channel system for analysis of theophylline. Furthermore, the authors studied the effect of applied voltage (0.09, 1.07, 1.05, 1.03, and 1.01 kV/cm) on the separation of FITC-arginine amino acid, which is shown in Fig. 7.19, indicating low limit of detection at

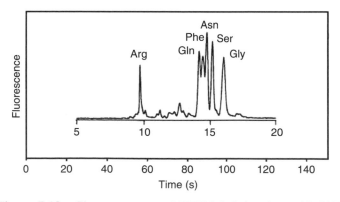

Figure 7.18   Chromatograms of FITC-labeled amino acids [42].

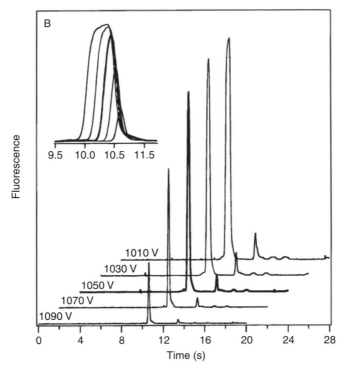

Figure 7.19   Effect of voltage on the height (limit of detection) of FITC-arg. amino acid [42].

1.01 kV/cm. Culbertson et al. [43] optimized the separation of 19 TRITC-labeled amino acids in a 10 mM sodium tetraborate-50 mM SDS buffer with 20% (v/v) methanol and 10% 2-propanol separately. The field strength was 770 V/cm, and the detection point was 11.87 cm from the injection cross. The best separation was achieved with 20% methanol as the modifier with quite good values of separation factors. Roman et al. [44] reported the separation of 6-carboxytetramethylrhodamine, succinimidyl ester-labeled amino acids and BODIPY FL CASE (*N*-(4,4-difluoro-5,7-dimethyl-4-bora-3a,4a-diaza-s-indacene-3-propionyl) cysteic acid, succinimidyl ester)-labeled amino acids. The mobile phase used was 10 mM sodium tetraborate and 20% acetonitrile with 25 mM SDS and 650 V/cm as applied voltage. Suljak et al. [45] described nanomicellar electrokinetic chromatography with electrochemical detection for the separation of many cationic catecholamines, such as dopamine, norepinephrine, and epinephrine. Furthermore, resolution was enhanced by coupling small internal diameter (5 μm) sampling capillaries with submicrometer internal height separation channels.

Ceriotti et al. [46] described nanomicellar electrokinetic chromatography on a microchip for the separation of low and high density lipoproteins (LDL and HDL). The LDL peak showed a focusing effect and exhibited an apparent efficiency of 2.2×10^7 plates/meter theoretical plates by using SDS as micelle formation agent. Furthermore, it was reported that low concentration of SDS did not significantly alter lipoprotein particle size; distribution within the time course. Peak sharpening effect was observed only when SDS was added slowly to the sample, probably due to a mobility gradient created between sample and running buffer. Shadpour and Soper [47] reported two-dimensional separation of a protein mixture on a PMMA microchip with 12 mM Tris-HCl (0.4% w/v, 14 mM) containing 0.05% w/v SDS (pH 8.5), with methylhydroxyethylcelullose (MHEC) as the dynamic EOF suppressor. Roman et al. [44] described separation of AlexaFluor 488-labeled *Escherichia coli* bacterial homogenates on PDMS chips. The mobile phase used was 10 mM sodium tetraborate and 20% acetonitrile with 25 mM SDS.

7.3.2 Environmental Samples

Normally, NMEKC requires only small amounts of sample and mobile phase and, hence, is capable of analyzing low amounts of samples with low limits of detection. Therefore, it may be used to identify various pollutants, which are present at trace levels. Some examples of NMEKC in environmental analyses are discussed here.

Moore et al. [48] used NMEKC for the separation of three coumarin dyes (coumarin 440, coumarin 450, and coumarin 460) on glass microchips. The separation capillary was 16.5 cm long for a serpentine channel chip and 1.3 cm long for a straight channel chip. The mobile phase used was buffer composed of 10 mM sodium borate (pH 9.1), 50 mM SDS, and 10% (v/v) methanol with >400 V/cm as applied voltage. Detection of analyte zones was accomplished by laser induced fluorescence detection (Fig. 7.20). Ramsey and Collins [49] discussed design, fabrication, and autonomous operation of an integrated microfluidic device for solid phase extraction coupled to NMEKC separation of rhodamine B at 60 fM detection limit. The authors used porous plugs of polymethacrylate polymer for fabricating the microchannels. A sample of dye was concentrated by SPE, eluted in a nonaqueous solvent from a separate on-chip reservoir, and injected by a gated valve onto a separate column for NMEKC analysis. The authors reported a completely automated sequence of extraction, elution, injection, separation, and detection in less than 5 min (Fig. 7.21). Roman et al. [44] reported a simple method for the effective and rapid separation of hydrophobic molecules (rhodamine B) by using NMEKC on PDMS. SDS served two critical roles, that is, it provided a

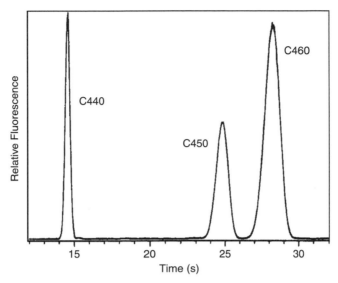

Figure 7.20   Chromatogram of the separation of coumarin dyes (C440, C450, and C460) [48].

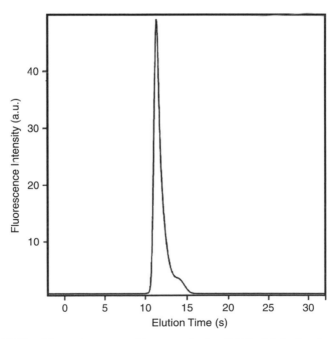

Figure 7.21   Chromatograms of trace analysis of 100 pM rhodamine B [49].

dynamic coating on the channel wall surfaces and formed a pseudo-stationary chromatographic phase. The SDS coating generated an EOF of 7.1×10^{-4} $cm^2 V^{-1}s^{-1}$ (1.6% RSD, $n = 5$), and eliminated the absorption of rhodamine B into the bulk PDMS with experimental conditions as mentioned above.

Hilmi et al. [50] reported the electrochemical reduction of trinitrotoluene (TNT), octahydro-1,3,5,7-tetranitro-1,3,5,7-tetrazocine (HMX), hexahydro-1,3,5-trinitro-1,3,5-triazine (RDX), and 10 other explosives by NMEKC. The mobile phase used was borate buffer (15 mM, pH 8.7) containing 25 mM SDS with 20 kV/cm as applied voltage. The detection mode was ampero-metric and the reported method was also used to determine the explosive con-tents of soil extracts and groundwater. Wallenborg and Bailey [51] described separation and detection of explosives using NMEKC and indirect laser-induced fluorescence. The detection set up used was in an epifluorescence configuration with excitation provided by a near-IR diode laser operating at 750 nm. The explosives studied were trinitrobenzene (TNB), trinitrotoluene TNT, dinitrobenzene (DNB), tetryl- and 2,4-dinitrotoluene (2,4-DNT). Analysis was completed within 60 seconds and 60,000 theoretical plates per meter were observed. The mobile phase used was borate buffer (50 mM, pH 8.5) having 50 mM SDS, with 4.0 kV/cm as separation voltage. The authors

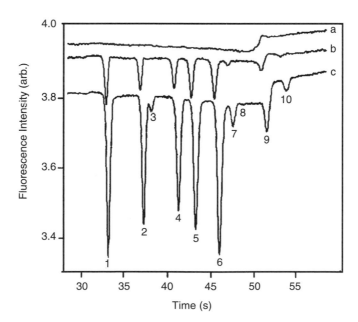

Figure 7.22 Chromatograms of (1) TNB, (2) DNB, (3) NB, (4) TNT, (5) tetryl, (6) 2,4-DNT, (7) 2,6-DNT, (8) 2-,3,4-NT, (9) 2-Am-4,6-DNT, and (10) 4-Am-2,6-DNT after extraction from (a) unspiked soil, (b) 1.0 ppm, and (c) 5.0 ppm spiked soils [51].

also reported the separation of two nitramines (HMX and RDX). Furthermore, on-chip injection schemes were evaluated by the authors for increasing sensitivity. A 250 μm double-T injector resulted in 35% increase in peak signal compared to a straight-cross injector. Figure 7.22 represents NMEKC-indirect laser-induced fluorescence (NMEKC-IDLIF) analysis of TNB, DNB, NB, TNT, tetryl, 2,4-DNT, 2,6-DNT, 2-,3,4-NT, 2-Am-4,6-DNT and 4-Am-2,6-DNT from extracts of blank soil and spiked with 1.0 and 5.0 ppm concentrations of each analyte. Wakida et al. [52] reported NMEKC separation of 4-nonylphenol, 4-(1,1,3,3-tetramethylbutyl)phenol, and bisphenol A in water. The authors combined this with a UV detector and achieved analysis within 15 seconds. The RSD values for peak height in 50 mg/L phenolic chemicals were less than 8% except for bisphenol A (11.0%). The method was used to analyze these pollutants in rivers, lakes, sea and ground waters.

7.3.3 Mechanisms of Separation

Basically, a surfactant is a molecule possessing two zones of different polarities, which shows special characteristics in solution. A surfactant in nanomicellar chromatography has a long chain hydrocarbon tail and charged head. The formation of micelles occurs in aqueous mobile phase when the concentration of the counterions exceeds a critical micelle concentration. Approximately 40 to 100 ions aggregate to form roughly spherical particles, with the hydrophobic tail oriented towards the center and the polar head pointed towards the outside of the micellar particle. Surfactants are divided into three categories, ionic (cationic and anionic), nonionic, and zwitterionic. The most important properties of surfactants are their critical micellar concentration (CMC), aggregation number, and Kraft point. CMC depends on temperature, salt concentration, and buffer additives. The aggregation number indicates the number of surfactant molecules taking part in micelle formation. The Kraft point is the temperature above which the solubility of the surfactant increases steeply due to the formation of micelles. The best surfactant for NMEKC possesses a good solubility in a buffer solution, forming a homogeneous micellar solution that is compatible with the detector and has a low viscosity.

Surfactant molecules containing long alkyl chain (hydrophobic group) and charged or neutral head (polar group) aggregate in aqueous solutions above their critical micellar concentration (CMC) and form micelles. These types of aggregates are spherical in shape with polar and hydrophobic groups at the outer and core regions, respectively. This micelle works as a pseudo-stationary phase that possesses a self-mobility different to that of the surrounding aqueous phase. Accordingly, the micellar phase acts as the stationary phase while the aqueous phase acts as the mobile phase in NMEKC. The distribution of species occurs between the micellar and aqueous phases. The solute and micellar interactions are of three types: (1) the solute is adsorbed on the surface of the micelle by electrostatic or dipole interactions, (2) the solute behaves as a co-surfactant by participating in the formation of the micelle, and (3) the solute is incorporated into the core of the micelle [53]. The extent of these interactions depends on the analytes and the micelle. Highly polar species are mainly adsorbed on the surface of the micelle and low polar solutes interact at the core of the micelle. Basically, the micelle and buffer tend to move towards the positive and negative ends, respectively. The movement of the buffer is stronger than the micelle movement and as a result the micelle and the buffers move toward the negative end. During separation of analytes, a second phase is formed and analytes are solubilized into the micelle. The species with different polarities are partitioned between the aqueous phase, stationary phase, micellar hydrophobic phase, and micellar hydrophilic phase and, hence, analytes are separated easily.

7.4. CONCLUSION

In spite of the wide range applications of NLC and NCE, NCEC and NMEKC are being developed due to their special inherent characteristic features as mentioned above, and required for specific purposes. The increase in number of publications during the last decade supports these facts. These modalities of liquid chromatography are still in development stages. Hopefully, these techniques will be highly useful in the coming years in many areas of nanoscience.

REFERENCES

1. T. Tsuda, K. Nomura, G. Nakagawa, *J. Chromatogr.*, **248**, 241 (1982).
2. J.H. Knox, I.H. Grant, *Chromatographia*, **24**, 135 (1987).
3. A.S. Cohen, A. Paulus, B.L. Karger, *Chromatographia*, **24**, 15 (1987).
4. V. Pretorius, B.J. Hopkins, J.D. Schieke, *J. Chromatogr.*, **99**, 23 (1974).
5. S.C. Jacobson, R. Hergenroder, L.B. Koutny, J.M. Ramsey, *Anal. Chem.*, **66**, 2369 (1994).
6. T.B. Stachowiak, F. Svec, J.M. Fréchet, *J. Chromatogr. A*, **1044**, 97 (2004).
7. I. Miksík, P. Sedláková, *J. Separation Sci.*, **30**, 1686 (2007).
8. I.M. Lazar, L. Li, Y. Yang, B.L. Karger, *Electrophoresis*, **24**, 3655 (2003).
9. M. Pumera, *Talanta*, **66**, 1048 (2005).
10. M. Pumera, *Electrophoresis*, **28**, 2113 (2007).

11. I. Nischang, U. Tallarek, *Electrophoresis*, **28**, 611 (2007).

12. K. Huikko, R. Kostiainen, T. Kotiaho, *Eur. J. Pharm. Sci.*, **20**, 149 (2003).

13. G.J.M. Bruin, *Electrophoresis*, **21**, 3931 (2000).

14. A. de Mello, *Lab. Chip*, **2**, 48N (2002).

15. N. Lion, T.C. Rohner, L. Dayon, I.L. Arnaud, E. Damoc, N. Youhnovski, Z.Y. Wu, C. Roussel, J. Josserand, H. Jensen, J.S. Rossier, M. Przybylski, H.H. Girault, *Electrophoresis*, **24**, 3533 (2003).

16. J.P. Kutter, *Trends Anal. Chem.*, **19**, 352 (2000).

17. V. Kašicka, *Electrophoresis*, **24**, 4013 (2003).

18. F.E. Regnier, B. He, S. Lin, J. Busse, *Trends Biotechnol.*, **17**, 101 (1999).

19. K. Mistry, I. Krull, N. Grinberg, *J. Separation Sci.*, **25**, 935 (2002).

20. L. Ceriotti, E. Verpoorte, *Anal. Chem.*, **74**, 639 (2002).

21. J. Park, D. Lee, W. Kim, S. Horiike, T. Nishimoto, S.H. Lee, C.H. Ahn, *Anal. Chem.*, **79**, 3214 (2007).

22. K. Faure, M. Blas, O. Yassine, N. Delaunay, G. Crétier, M. Albert, J.L. Rocca, *Electrophoresis*, **28**, 1668 (2007).

23. B.E. Slentz, N.A. Penner, E. Lugowska, F. Regnier, *Electrophoresis*, **22**, 3736 (2001).

24. B.E. Slentz, N.A. Penner, F.E. Regnier, *J. Chromatogr. A*, **948**, 225 (2002).

25. B.E. Slentz, N.A. Penner, F.E. Regnier, *J. Separation Sci.*, **25**, 1011 (2002).

26. N. Gottschlich, S.C. Jacobson, C.T. Culbertson, J.M. Ramsey, *Anal. Chem.*, **73**, 2669 (2001).

27. D.J. Throckmorton, T.J. Shepodd, A.K. Singh, *Anal. Chem.*, **74**, 784 (2002).

28. R. Jindal, S.M. Cramer, *J. Chromatogr. A*, **1044**, 277 (2004).

29. M. Galloway, S.A. Soper, *Electrophoresis*, **23**, 3760 (2002).

30. M. Galloway, W. Stryjewski, A. Henry, S.M. Ford, S. Llopis, R.L. McCareley, S.A. Soper, *Anal. Chem.*, **74**, 2407 (2002).

31. J.P. Kutter, S.C. Jacobson, N. Matsubara, J.M. Ramsey, *Anal. Chem.*, **70**, 3291 (1998).

32. Y. Fintschenko, W.Y. Choi, S.M. Ngola, T.J. Shepodd, *Fresen. J. Anal. Chem.*, **371**, 174 (2001).

33. C. Ericson, J. Holm, T. Ericson, S. Hjertén, *Anal. Chem.*, **72**, 81 (2000).

34. M. Pumera, J. Wang, E. Gruska, R. Polsky, *Anal. Chem.*, **73**, 5625 (2001).

35. B.S. Broyles, S.C. Jacobson, J.M. Ramsey, *Anal. Chem.*, **75**, 2761 (2003).

36. Y. Wang, Z. Zhang, L. Zhang, F. Li, L. Chen, Q.H. Wan, *Anal. Chem.*, **79**, 5082 (2007).

37. S. Hjertén, L. Valtcheva, K. Elenbring, J.L. Liao, *Electrophoresis*, **16**, 584 (1995).

38. S. Terabe, K. Otsuka, A. Ichikawa, T. Ando, *Anal. Chem.*, **56**, 111 (1984).

39. K. Huikko, R. Kostiainen, T. Kotiaho, *Europ. J. Pharm. Sci.*, **20**, 149 (2003).

40. V. Dolnik, S. Liu, *J. Separation Sci.*, **28**, 1994 (2005).

41. M. Pumera, *Electrophoresis*, **27**, 244 (2006).

42. F. von Heeren, E. Verpoorte, A. Manz, W. Thormann, *Anal. Chem.*, **68**, 2044 (1996).

43. C.T. Culbertson, S.C. Jacobson, J.M. Ramsey, *Anal. Chem.*, **72**, 5814 (2000).

44. G.T. Roman, K. McDaniel, C.T. Culbertson, *Analyst*, **131**, 194 (2006).

45. S.W. Suljak, L.A. Thompson, A.G. Ewing, *J. Separation Sci.*, **27**, 13 (2004).

46. L. Ceriotti, T. Shibata, B. Folmer, B.H. Weiller, M.A. Roberts, N.F. de Rooij, E. Verpoorte, *Electrophoresis*, **23**, 3615 (2002).

47. H. Shadpour, S.A. Soper, *Anal. Chem.*, **78**, 3519 (2006).

48. A.W. Moore, Jr., S.C. Jacobson, J.M. Ramsey, *Anal. Chem.*, **67**, 4184 (1995).

49. J.D. Ramsey, G.E. Collins, *Anal. Chem.*, **77**, 6664 (2005).

50. A. Hilmi, J.H.T. Luong, A.L. Nguyen, *Anal. Chem.*, **71**, 873 (1999).

51. S.R. Wallenborg, C.G. Bailey, *Anal. Chem.*, **72**, 1872 (2000).

52. S. Wakida, K. Fujimoto, H. Nagai, T. Miyado, Y. Shibutani, S. Takeda, *J. Chromatogr. A*, **1109**, 179 (2006).

53. S. Terabe, *J. Pharm. Biomed. Anal.*, **10**, 705 (1992).

CHAPTER 8

NANOCAPILLARY ELECTROPHORESIS

8.1. INTRODUCTION

The idea of modern electrophoresis originated from the experiments of Kohlrausch [1] on the migration of ions in an electrolyte solution. Later on Arne Tiselius [2] separated protein mixtures by electrophoresis, which won him the Nobel prize in 1948. The first capillary electrophoresis apparatus was designed by Hjerten in 1967 [3] and the modern era of CE is considered to have begun with the publications of Jorgenson and Lukacs [4–6] describing CE instrumentation. Although CE had been a topic of discussion among scientists it gained recognition in 1989 during the First International Symposium on High Performance Capillary Electrophoresis, held in Boston [7]. This meeting attracted over 400 scientists to attend 100 presentations on the theory and practice of CE.

Initially, electrophoresis was performed in gel or other media in the form of a bed, slab, rod, etc., but due to laborious multistage handling of supporting media and nonreproducibility of the results the supporting medium was replaced by a capillary and the technique was called capillary electrophoresis (CE). More recently, the silica capillary was replaced by a microchip and the technique is called microchip electrophoresis; we define it as nanocapillary electrophoresis (NCE) as it deals with all aspects at nano or low levels of

Nanochromatography and Nanocapillary Electrophoresis. By Ali, Aboul-Enein, and Gupta
Copyright © 2009 John Wiley & Sons, Inc.

quantitation. It has been realized that NCE is a versatile technique of high speed, high sensitivity, and inexpensive running cost, which is an innovation in separation science. During the last few years, NCE has been used for qualitative and quantitative separations in proteomics, genomics, drug development, and design and environmental analyses at nano or low levels of detection. Reviews have been published in the literature on NCE [8–22]. This chapter describes the state of the art of NCE for the separation and identification of different compounds. The topic includes optimization, applications, and mechanism of separation.

8.2. OPTIMIZATION

As in the case of conventional CE, optimization is also the most important aspect in NCE. Optimization is the first step in the development of a method for any NCE analysis. The most important parameter to be controlled is choice of buffer, including concentration and pH. Applied voltage and amount of injection are also important for optimization. For the best separation a moderate voltage is maintained in NCE using buffers as the background electrolyte (BGE). The conductivity of the BGE should be higher than the conductivity of the sample, which can be obtained by using buffers as the BGE. Besides, buffers are useful to control pH of the BGE throughout the experiments. Therefore, buffers are used as BGEs in most NCE applications. The most commonly used buffers are phosphate, acetate, borate, ammonium citrate, Tris, among others that are used with different concentrations and pHs. The electrolyte identity and concentration must be chosen carefully for optimum analysis.

The selection of the BGEs depends on their conductivity and the type of species to be studied. The relative conductivities of different electrolytes can be estimated from their condosities (defined as the concentration of sodium chloride, which has the same electrical conductance as the substance under study). Low UV absorbing components are required for the preparation of the buffers, if the detection is carried out by UV detector. Most of the applications of NCE are performed using techniques with MS hyphenation and, hence, volatile components are required for high sensitivity and long life of the machine. These conditions substantially limit the choice to a moderate number of electrolytes. pH of the BGE is also another factor that determines the choice of the buffers. For low pH buffers, phosphate and citrate have commonly been used although the later absorb strongly at wavelengths <260 nm. Basic buffers such as borate, Tris, CAPS, etc., are used as suitable BGEs.

Fang et al. [23] described analyses of fluorescein isothiocyanate (FITC) labeled amino acids, that is, arginine, phenylalanine, and glycine, in sodium tetraborate buffer (pH 9.2) within 8 to 80 seconds (Fig. 8.1). The authors optimized separations by varying sequential injection, buffer flow rate, and voltage, as shown in Figs. 8.2, 8.3, and 8.4. A perusal of these figures indicates a good separation at 80 µL injected volume, 1.0 mL/min flow rate, and 6.0 kV/cm as applied potential. Wang and coworkers [24] described NCE optimization of paraoxon, methyl parathion, fenitrothion, and ethyl parathion. The optimiziation was carried out using a MES buffer as BGE, a 72 mm long separation channel, and an applied voltage. The effects of separation voltage, running buffer, pH, and SDS concentrations are shown in Figs. 8.5, 8.6, 8.7, and 8.8 respectively. These figures indicate the best

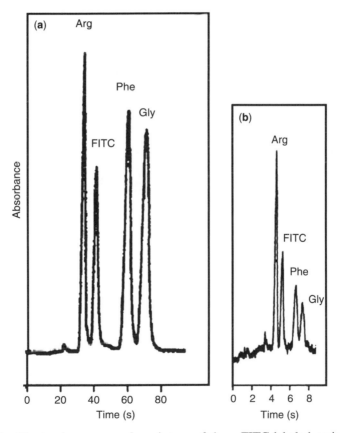

Figure 8.1 Electropherograms of a mixture of three FITC-labeled amino acids at (a) 25 and (b) 35 mm separation lengths [23].

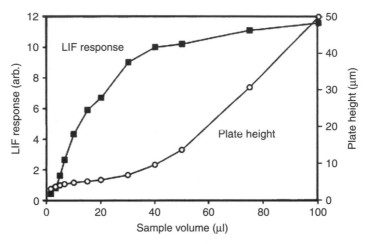

Figure 8.2 Effects of sequential injection sample volume on peak height and plate height of FITC-labeled amino acids [23].

separations at 155 to 2500 V, MES buffer (20 mM, pH 5.0) with 10 mM SDS concentration. The method was applied to spiked river water samples and the implications for on-site environmental monitoring and rapid security screening/warning were discussed. Wang and Chatrathi [25] optimized a novel NCE for the analyses of renal markers, that is, creatine, creatinine, p-amino-hippuric acid, and uric acid, in urine samples. The analysis was completed within 5 minutes by optimizing separation voltage and detection potential.

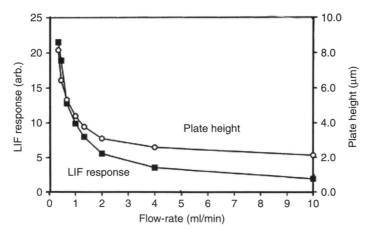

Figure 8.3 Effects of BGE flow rate on peak height and plate height of FITC-labeled amino acids [23].

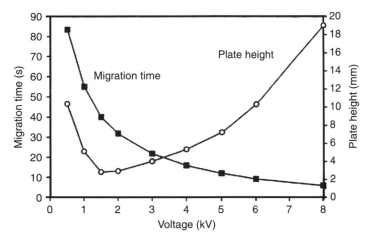

Figure 8.4 Effects of voltage on migration time and plate height of FITC-labeled amino acids [23].

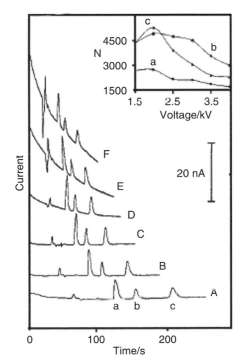

Figure 8.5 Effect of separation voltage on the response for a mixture containing (a) 3.0×10^{-5} M paraoxon, (b) 3.0×10^{-5} M methyl parathion, and (c) 6.0×10^{-5} M fenitrothion. Separation performed using (A) + 1500, (B) + 2000, (C) + 2500, (D) + 3000, (E) + 3500 and (F) + 4000 V, respectively. Inset: graph indicates the plate number (N) versus separation voltage [24].

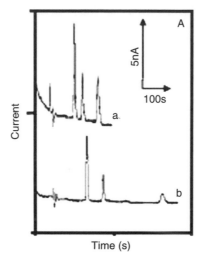

Figure 8.6   Effect of buffers. (a) MES and (b) acetate (20 mM, pH 5.0), containing 10 mM SDS on the separation of 7.1×10^{-5} M paraoxon, 7.5×10^{-5} M methyl parathion, and 1.4×10^{-5} M fenitrothion [24].

Jackson et al. [26] set miniaturized, battery-powered, high-voltage power supply conditions in NCE for the separation of dopamine and catecholamines with electrochemical detection. The authors varied platinum working electrode voltage from 25 to 200 V/cm for achieving low limits of detection.

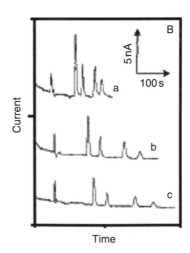

Figure 8.7   Effect of pH on MES buffers (20 mM), containing 10 mM SDS, of (a) pH 5.0, (b) 6.0, and (c) 7.0 on the separation of 7.1×10^{-5} M paraoxon, 7.5×10^{-5} M methyl parathion, and 1.4×10^{-5} M fenitrothion [24].

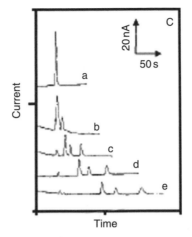

Figure 8.8 Effect of SDS concentrations (a) 0.0, (b) 2.5, (c) 5.0, (d) 7.5, and (e) 10 mM in MES buffers (20 mM, pH 5.0) on the separation of 7.1×10^{-5} M paraoxon, 7.5×10^{-5} M methyl parathion, and 1.4×10^{-5} M fenitrothion [24].

The optimized separation is shown in Fig. 8.9 indicating the best peak shape and low detection limits at 100 V/cm.

Smith et al. [27] described DNA separations in NCE with automated capillary sample introduction and laser-induced fluorescence (LIF) detection within

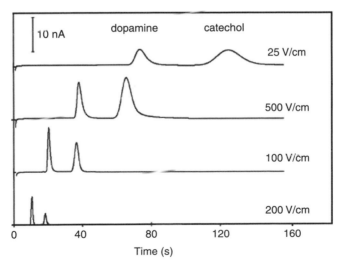

Figure 8.9 Separations of 2.2 mM dopamine and 4.6 mM catechol in 20 mM phosphate buffer, pH 6 [26].

60 seconds. Furthermore, about 550 separations of oligonucleotides could be performed in one hour by increasing the number of lanes to 37 and optimizing the rate of the manipulator movement. NCE sample introduction into chips allowed parallel separations to be continuously performed in serial, yielding high throughput and minimal need for operator intervention. Kikura-Hanajiri et al. [28] reported indirect measurement of nitric oxide production by monitoring nitrate and nitrite using NCE with electrochemical detection (ECD). The optimization of the separation, injection, detection, and reduction reaction conditions was described by the authors. Furthermore, the authors reported that the reduction reaction was successfully integrated on chip and carried out in approximately 60 seconds followed by activation of cadmium granules. The usefulness of this device was demonstrated by monitoring the amount of nitrate and nitrite produced from 3-morpholinosydnonimine, a NO-releasing compound. Guihen and Glennon [29] discussed rapid separation of antimicrobial metabolites by NCE with UV linear imaging detection, that is, monoacetylphloroglucinol (MAPG) and 2,4-diacetylphloroglucinol (2,4-DAPG) from *Pseudomonas fluorescens* F113. The separation was achieved on a separation channel length of 25 mm within 20 seconds. The authors optimized separation by sample introduction/injection parameters.

Ping et al. [30] developed a fast method for analysis of lipoproteins by microchip electrophoresis with light-emitting diode confocal fluorescence detection. Sodium dodecyl sulfate and cetyltrimethylammonium bromide were utilized to alter lipoproteins and channel surface for optimization. The peak shape of lipoproteins was greatly improved, demonstrating lipoprotein adsorption on a poly-(methyl methacrylate) (PMMA) chip; reducing electrostatic repulsion. The different separation parameters including surfactant concentration, buffer pH, and polymer concentration, as well as online concentration were investigated. Under optimal conditions, two baseline separations of standard lipoproteins, including high density lipoprotein, low density lipoprotein, and very low density lipoprotein were achieved with different selectivities. According to the authors, the method afforded high separation speed within 100 seconds and high reproducibility with intraassay and interassay relative standard deviations (RSDs) of lipoprotein migration times in the range of 0.90% to 1.9%, indicating the reliability of the method. Sikanen et al. [31] fabricated a NCE-ESI-MS device and used it for testing verapamil as a test compound. According to the authors the sheath flow interface enabled comprehensive optimization of both NCE and ESI conditions. Various radial NCEs for DNA sequencing consisting of 16, 96, and 384 channels were fabricated on 6 inch glass wafers [32–37]. One of them is shown in Fig. 8.10. Turn geometry has been studied both theoretically and

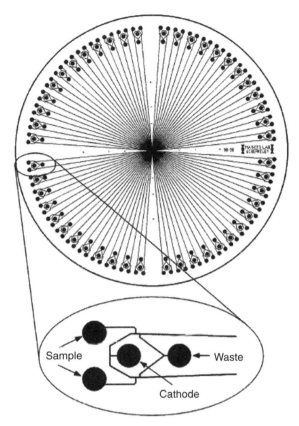

Figure 8.10   Diagram of a microfabricated DNA sequencing device with 96 channels fabricated on a single substrate of 6 inch in diameter [32].

experimentally to reduce turn dispersion and minimize the turn-induced resolution loss. Several parameters, including the radius of curvature of the turn, the tapering length, and the degree of tapering were optimized. High resolution DNA separation can be achieved in channels with a small radius of curvature ($\sim$250 μm), a short tapering length ($\sim$55 μm), and a large tapering ratio (4 : 1 separation channel width to turn channel width) [38]. Omasu et al. [39] coated NCE chips with three reagents, namely bovine serum albumin (BSA), gelatin, and 2-methacryloyloxyethylphosphorylcholine (MPC) polymer, and experiments were carried out in phosphate buffer (pH 4 to 9) using erythrocytes extracted from sheep whole blood. The electroosmotic flow (EOF) mobility was measured using noncharged particles and the effect of pH on electromigration in sheep erythrocyte was studied as shown in Fig. 8.11, which indicates high electrophoretic migration at low pH.

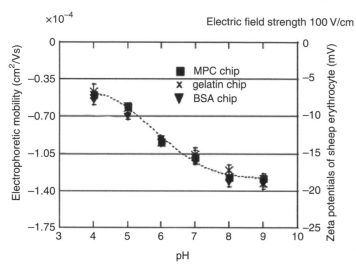

Figure 8.11 Electrophoretic mobility and zeta potentials of sheep erythrocyte. (Left) Relationship between true electrophoretic mobility in sheep erythrocyte and pH using NCE chips coated with BSA, gelatin, and MPC polymer. (Right) Relationship between sheep erythrocyte zeta potentials and pH using NCE chips coated with BSA, gelatin, and MPC polymer [39].

8.3. APPLICATIONS

After optimization, application is the next step in NCE to perform in real samples. Within a decade NCE has gained momentum in separation science. It has been used for analyzing all types of samples of biological and environmental importance. Attempts have been made to describe the applications of NCE in the areas of proteomics, genomics, amino acids, drug development and design, enzymes, hormones, biological fluids, foods and beverages, viruses, bacteria, explosives, environmental samples, and other applications. These are described in the following sections.

8.3.1. Proteomics

In living organisms, each cell produces thousands of proteins and one set of these proteins is called a proteome. Their analysis is a tedious job. Moreover, the low amounts of proteins in a proteome makes this task challenging. Fortunately, the development of microfluidic devices including NCE is the best innovation of the last decade to solve such types of riddles. Many papers have been published on proteomics analysis using NCE; some examples are discussed here. For proteomic analysis, protein digestion plays an important role in the sample pretreatment process. Normally, proteins are digested and converted into small peptides and analyzed.

Colyer et al. [40] reported separation of human serum proteins IgG for mimicking γ zone, transferrin β zone, α-1-antitrypsin and albumin by using an NCE-LIF device. Buffer conditions were 100 mM borate with 2 mM lactate (pH 10.5). Chiem and Harrison [41] described NCE for analysis of monoclonal antibodies and theophylline in serum samples within 40 seconds. The detection limit of theophylline was 1.25 ng/mL. Buffer used was 0.05 M tricine adjusted to pH 8.0, which allowed for adequate separation (40,000 plates for theophylline; 1000 plates for theophylline-antibody complex and for human IgG) and gave reproducibility of migration times of 1% to 1.5% over 4-day periods, indicating minimal problems from adsorption in the uncoated chips. Rodriguez et al. [42] reported the analysis of FITC-anti-human IgG. Analysis was obtained within 60 minute on 6 cm long capillary from injection end to detector with electric field strength of 0.268 kV/cm. Linder et al. [43] described using poly(dimethylsiloxane)-glass chips for a heterogeneous competitive human serum immunoglobulin G immunoassay employing Cy5 human IgG as tracer and Cy3-mouse IgG as internal standard. Forrer et al. [44] described a chip-based gel electrophoresis method for the quantification of half-antibody species in IgG4 and their by- and degradation products.

Chiem and Harrison [45] reported analysis of bovine serum albumin (BSA) in diluted mouse ascites fluid by using NCE. Separations were performed in less than 60 seconds on chip. Furthermore, the potential applications of this device in affinity measurements were discussed. Hsiung et al. [46] described NCE for the separation and identification of various proteins, including BSA and β-casein with detection by coupling two light sources of different wavelengths to the two excitation optic fibers. Chen et al. [47] described protein analysis of Cy5-labeled bovine serum albumin (Cy5-BSA) and anti-BSA by NCE with flow-through sampling technique. The device used hydrodynamic pressure for driving sample flow and a gating voltage was applied to the electrophoretic channel on the microchip to control the sample loading for separation and to inhibit sample leakage, with detection at the nanogram level. Zhuang and coworkers [48] described a low temperature bonding method for microfabrication of quartz microfluidic chips, which had been used for analysis of serum lipoproteins with LIF. According to this study, low and high density lipoproteins in the serum were separated completely using tricine buffer with methylglucamine. Sikanen et al. [49] compared the performance of SU-8 (epoxy based photoresist) separation microdevices with glass chips for the separation of biologically active peptides with fluorescent detection. The authors used a variety of buffers for the separation. Plate heights of

2.4 to 5.9 μm were obtained for fluorescent-labeled BSA. The authors compared the performance of two types of chips and it was observed that the analytical performance of SU-8 microchips was good and fairly comparable to that of commercial glass chips as well as that of traditional capillary electrophoresis and chromatographic methods. Besides, lithography-based patterning of SU-8 enabled straightforward integration of multiple functions on a single chip, which favored NCE.

Liu et al. [50] reported the separation of proteins by NCE with noncovalent post-column labeling. The limits of detection for model proteins, α-lactalbumin, β-lactoglobulin A, and β-lactoglobulin B, were <0.5 pg. Yue et al. [51] described sample preparation and separation in NCE for proteomic applications, including separation of β-casein in peptide mixtures. Li et al. [52] identified peptides from membrane proteins using NCE-TOF-MS. The detection limit achieved ranged from 3.2 to 43.5 nM for different peptides. The analyses of protein digests were typically achieved in less than 1.5 minutes. Deyl et al. [53] compared the separation of protein complexes by using NCE and CE and reported fast analysis with low detection limit of the former technique. The CE was carried out on the chip at 100 V/cm. Tsai et al. [54] used native and sodium dodecyl sulfate capillary gel electrophoresis of proteins on a single microchip within 20 minutes. Musyimi et al. [55] used NCE-MALDI-TOF-MS for the separation of peptides and peptide fragments produced from a protein digest. The linear separation channel was 50 μm wide and 100 μm deep, and possessed an 8.0 cm effective length separation channel with a double-T injector (Vinj = 10 nL). The exit of the separation channel was machined for allowing direct contact deposition of effluent onto a specially constructed rotating ball inlet to MS. Matrix addition was accomplished in-line on the surface of the ball. The coupling utilized the ball as the cathode transfer electrode to transport sample into the vacuum for desorption.

Nagata et al. [56] described high-speed separation of proteins by NCE on a polyethylene glycol-coated plastic chip with a sodium dodecyl sulfate-linear polyacrylamide solution. According to the authors the electrophoretic separation of proteins (21.5 to 116.0 kDa) was completed with separation lengths of 3 mm, achieved within 8 seconds. Fruetel and coworkers [57] described a separation of fluorescamine-labeled protein biotoxins on a single fused-silica wafer containing two separation channels. The detection was achieved using a miniaturized LIF detector employing two diode lasers and one per separation channel. The authors tested the portability of this instrument in a laboratory field test at the Defense Science and Technology Laboratory with a series of biotoxin variants. Dahlin et al. [58] prepared hybrid capillary PDMS microchips integrated with ESI tips by casting PDMS in a mold. The separation of peptides was carried out within 2 minutes using the unit.

Sun et al. [59] used NCE-LIF to evaluate a derivatization method mediated by liposome for single cell analysis. According to these workers single cell analysis revealed that liposome-membrane fusion occurred after entrance of liposomes into the cells, with release of encapsulated fluorescence dyes and labeling of intracellular species.

Mohanty et al. [60] developed NCE for analysis of proteins. An electrical field of 40 V/cm was used to separate and collect the proteins. These workers reported this device was simple to fabricate, benefited from microscale analysis, and included an on-chip collection scheme that interfaced the macro world with the micro world. Chow [61] described the use of two types of commercialized microfluidic chips for protein separation, suitable for personal-scale and high throughput purposes. Silvertand et al. [62] described an on-line sample pretreatment and analysis of proteins and peptides on a poly(methylmethacrylate) (PMMA) microfluidic device (IonChip). The chip consisted two hyphenated electrophoresis channels with integrated conductivity detectors. The first channel was used for sample pre-concentration and sample clean-up, while the second channel was used for separation. Phillips and Wellner [63] developed two chip-based capillary electrophoresis systems for analyzing inflammatory neuropeptides in tissue fluids of patients with neuropeptide-associated muscle pain. The first and second chips were used to perform electrokinetic flow immunoassays and an immunoaffinity port containing an array of immobilized antibodies was used to capture to capture the analytes of interest. The authors reported that both chip-based systems provided a relatively fast, accurate procedure for studying inflammatory biomarkers in complex biological fluids. Furthermore, these workers analyzed 12 different inflammation-associated mediators in about 2 minutes.

Zhang et al. [64] used NCE-MS on glass wafers with standard photolithographic/wet chemical etching methods. The design integrated with sample inlet ports, the separation channel, a liquid junction, and a guiding channel for the insertion of the electrospray chamber of an ion trap MS. This device was used for analyses of peptides, proteins, and protein tryptic digests. High efficiency was obtained on a longer channel of 11 cm in the on-chip micro-device, with fast separations within 50 seconds. The samples were injected via both electrokinetic and pressure-driven forces. A typical electropherogram of the angiotensin peptides mixture is shown in Fig. 8.12. Furthermore, the authors reported separation and identification of 12 angiotensin peptide mixtures obtained from different animals on the microdevice with a 4.5 cm separation channel, with 20 μg/L as sample concentration of each peptide. The BGE used was 20 mM 6-aminocaproic acid/acetic acid with pH 4.4. The applied voltage was 650 V/cm. The expanded electropherograms of

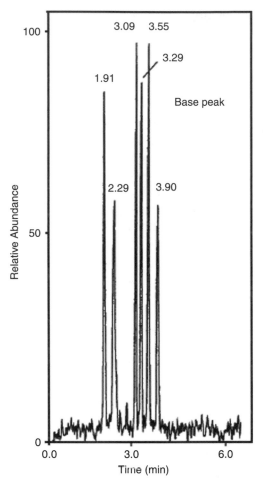

Figure 8.12   Electropherogram of angiotensin peptides mixture [64].

these 12 angiotensin peptides are given in Fig. 8.13. Vasilyeva et al. [65] used microfluidic technology for quantification of a half-antibody in immunoglobulin G4 samples. The authors referred to this method as a chip-based capillary gel electrophoresis (GelChip-CE method) in their work. Renzi et al. [66] designed, fabricated, and demonstrated a hand-held microchip-based analytical instrument for detection and identification of proteins and other biomolecules. The unit was called a μ-ChemLab, which had a modular design providing reliability and flexibility. The components included two independent separation modules, incorporating interchangeable fluid cartridges (2 cm^2 fused silica microfluidic chip) and a miniatured laser-induced fluorescence detection module.

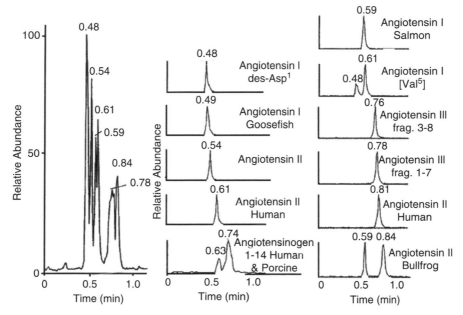

Figure 8.13 The expanded electropherograms of 12-angiotensin peptides on NCE with a 4.5 cm separation channel [64].

8.3.2. Genomics

NCE is a very useful technique for DNA analysis and has opened up new realms in genomics due to far reaching benefits with a decrease in the cost per analysis. The development of miniaturized genotyping platforms has supported breakthroughs in biomedical science. During the last few decades much work has been carried out in genomic research to map and sequence complete genomes of various animals and plants, including human beings. The amount of DNA is very small and, hence, microfluidic devices played a crucial role in purification and separation of DNA from the cell. NCE is of great importance in this concern and some workers have used this technique for genomic research. Kan et al. [67] and Lin et al. [68] reviewed NCE for DNA purification and sequencing. A radial microfabricated DNA sequencing device consisting of 96 channels fabricated on a 6 inch glass wafer was described by Shi et al. [32] and Paegel et al. [33] and is shown in Fig. 8.10. This sort of fabrication with other auxiliary tools such as a high-speed, rotary LIF scanner and a specialized, high-pressure sieving matrix loader are keys in enabling high throughput genetic analyses of DNA in a single electrophoresis device [69–73]. Other NCE devices comprised of 16 and 384 channels have been fabricated on 6 inch glass wafers [34–37,72]. In these devices, all the channels were fabricated on a single planar substrate providing the necessary separation length.

Meagher et al. [74] reported a bioconjugate approach for performing highly multiplexed single-base extension assays demonstrated by genotyping a large panel of point mutants in axons 5 to 9 of the p53 gene. The authors created a series of monodisperse polyamide drag-tags by using both chemical and biological synthesis. Furthermore, the authors achieved high resolution separation of genotyping reaction products by microchannel electrophoresis without a polymeric sieving matrix, with 70 seconds as separation time. Chowdhury et al. [75] described an NCE method to genotype for common single nucleotide polymorphisms in the thiopurine S-methyltransferase gene, which lead to serious adverse drug reactions for patients undergoing thiopurine therapy. The authors reported 100% concordance between NCE and conventional methods in a total of 80 patients.

Liu et al. [76] described an automated parallel DNA sequencing on multiple 16-channel microchips. Under computer control, high voltage was applied for sequencing the DNA samples. An integrated four-color confocal fluorescent detector was able to scan more than 450 bases in 15 minutes in all 16 channels. The advantages of DNA sequencing by the reported NCE were uniform signal intensity and tolerance of high DNA template concentration. Hong et al. [77] reported a microfabricated polymer chip of PDMS for NCE used for separation of different sizes of DNAs. Furthermore the same group [78] studied gene amplification in NCE (PDMS and glass). The method has been reported as an inexpensive single use apparatus. Tang et al. [79] described an isothermal signal amplification technique for specific DNA sequences, known as cycling probe technology (CPT) in an integrated NCE. An off-chip CPT reaction, with on-chip separation gave a detection limit of 2 fM target DNA and an amplification factor of 85,000. Xu et al. [80] described NCE analysis of DNA fragments using electrokinetic injection with transient isotachophoretic preconcentration. The channel of the microchip was 40.5 mm long, and 110 μm wide, and 50 μm deep. Liu et al. [81] described isotachophoresis nanochip electrophoresis (ITP-ZE) for hepatitis B virus (HBV) genotyping test in clinical diagnosis.

Kataoka et al. [82] described the usefulness of NCE (Hitachi model SV 1100 microchip) for analyzing nonstandard DNA samples. The analysis was completed within 4 minutes, with a detection limit of 1.83 ng/μL. Lin et al. [83] described a novel NCE system on a glass chip for separation and identification of DNA in a sample. Furthermore, the authors reported that this system was capable of highly efficient separations of other biomolecules. Obeid et al. [84] differentiated genetically modified organisms (GMOs) using NCE-LIF. The chip was composed of two glass plates, each 25×76 μm, thermally bonded together to form a closed structure. Posedi et al. [85] used NCE for the differentiation of the closely related cyathostomin species *Cylicocyclus elongatus* and *Cylicocyclus insigne* from the horse, based on genetic comparison. Furthermore, the authors indicated that the procedure described has provided an additional powerful tool, which might be employed for species delineation of closely related strains or species, such as the two taxa of *Cylicocyclus*. Similarly, Dooley et al. [86] used NCE and determined the mitochondrial cytochrome b genes of fish in the United Kingdom by which a differentiation of fish species was carried out. This method was used by U.K. Food Control Laboratories for product identification. Spaniolas et al. [87] described a molecular genetics approach to differentiate *Arabica* and *Robusta* coffee beans by using NCE. The plastid copy number was relatively constant across a wide range of bean samples, suggesting the methodology was useful for the quantification of any adulteration of *Arabica* with *Robusta* beans.

Fu and Lin [88] described high-resolution DNA separation via NCE by utilizing double-L injection techniques. The experimental and simulation results indicated that the unique injection system employed in the current microfluidic chip had the ability to replicate the functions of both the conventional cross-channel and the shift-channel injection systems. The authors advocated this system as highly useful in high resolution, high throughput biochemical analyses in many other areas. Xu et al. [89] reported chip gel electrophoresis for high sensitivity detection of DNA by combining electrokinetic injection with transient isotachophoresis preconcentration. Chen and Burns [90] described the effect of buffer flow on DNA separation in a microfabricated electrophoresis system. The authors also investigated the effect of buffer concentration on resolution and no improvement could be obtained by increasing the buffer concentration without flow. Sieben and Backhouse [91] described labeling of DNA along with high separations of DNA by using NCE. Furthermore, the authors reported a method of controllably labeling DNA fragments at the end of the electrophoretic separation channel in a glass microfluidic chip. Hawtin et al. [92] described the analyses of nucleic acid by NCE. The authors also reported an improvement in workflow processes, speed of analysis, data accuracy and reproducibility, and automated data analysis. Kawabata et al. [93] described liquid-phase binding assay of α-fetoprotein using DNA-coupled antibody and NCE. The immunoassay method, utilizing a liquid-phase binding assay format, was simple and convenient for antigen measurements on microchips.

Kim and Kang [94] described on-channel base stacking in NCE for high sensitivity DNA fragment analysis. Chiesl et al. [95] used NCE for purification of DNA. The octylacrylamide and dihexylacrylamide copolymers were used as adsorbents for proteins from DNA solution. Chuang et al. [96] described the binding of estrogen receptor (ER) to estrogen response element (ERE) in genomic pathways of estrogens by analyzing gel-based electrophoretic mobility shift assay (EMSA) on microchip electrophoresis using PEG-modified glass microchannels, which bear neutral surfaces against the adsorption of acidic DNA molecules and basic ER proteins. Furthermore, the authors demonstrated the feasibility of their method by measuring binding constants of recombinant ERalpha and ERbeta with a consensus ERE sequence (cERE, 5'-GGTCAGAGTGACC-3') as well as with an ERE-like sequence (ERE 1576, 5'-GACCGGTCAGCGGACTCAC-3').

Blazej et al. [97] described an efficient, nanoliter-scale microfabricated bioprocessor integrating all three Sanger sequencing steps, thermal cycling, sample purification, and capillary electrophoresis on a hybrid glass PDMS chip. NCE integration enabled complete Sanger sequencing from only 1 fmol of DNA template. The authors have sequenced up to 556 continuous bases with 99% accuracy, demonstrating read lengths required for de novo sequencing of human and other complex genomes. The authors reported the performance of this miniaturized DNA sequencer has provided a benchmark for predicting the ultimate cost and efficiency limits of Sanger sequencing.

Huang et al. [98] developed PMMA and polycarbonate-based chips for electrophoresis and tested their functionality by separating *phiX*174 DNA/ *Hae*III markers. According to the authors the experimental data showed that S/N chips using PE/TPE film was 5.34, when utilizing DNA markers with a concentration of 2 ng/μL with buffer of 2% hydroxypropyl-methylcellulose (HPMC) in Tris-borate-EDTA (TBE) with 1% YO-PRO-1 fluorescent dye. Furthermore, a mixture of an amplified antibiotic gene of *Streptococcus pneumoniae* and *phiX*174 DNA/*Hae*III markers was successfully separated and detected by using the developed chips. It has also been reported that DNA samples were separated within 2 minutes.

Among many miniaturized analytical devices, the polymerase chain reaction (PCR) microchip and microdevices have been studied extensively in various fields, with special emphasis on DNA purification and sequencing. The PCR reaction vessels can increase resolution while reducing the overall size of the PCR device with the miniaturization of PCR amplification needed to match NCE. Dolnik and Liu [99] and Zhang et al. [100] presented reviews on fabrication (bonding and sealing) of PCR microfluidic devices and their applications in NCE. Legendre et al. [101] described online SPE-NCE for DNA isolation from an anthrax spore-spiked nasal swab through PCR. Nojima et al. [102] described NCE for the isolation of target DNA species from a DNA mixture generated by PCR, whose starting material was a ligation mixture of an insert and an expression vector. The authors reported that the total operation in standard genetic engineering could be performed in a cell-free condition with the help of NCE. Lutz-Bonengel et al. [103] described NCE for low volume amplification for analyses of nuclear DNA. Qin et al. [104] described NCE coupled with PCR restriction fragment length polymorphism (RFLP) assay for genotyping A (-6) G single nucleotide polymorphism (of the angiotensinogen [AGT] gene, the candidate gene for essential hypertension) in 123 patients. The separation and detection of the digested PCR amplicons were completed in just 280 seconds. The genotyping of the A (-6) G polymorphism of the AGT gene in the core promoter region is shown in Fig. 8.14.

The main focus of functional genomic studies in the post-genomic era is the analysis of gene variants in human beings for disease diagnosis, prognosis, and management. NCE has also been used to recognize mutation in the tumor susceptibility genes BRCA1 and BRCA2 [105]. The separation in a 3 cm long channel took less than 120 seconds. Buch et al. [106] described a polymer microfluidic system for the detection of DNA point mutations via

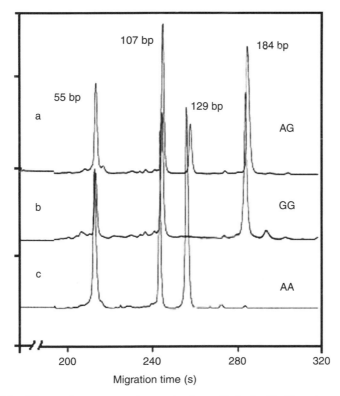

Figure 8.14 Electropherogram of genotyping of A (-6) G polymorphism of AGT gene, (a) A/G heterozygote (55 bp, 107 bp, 129 bp, 184 bp), (b) G/G homozygote (55 bp, 107 bp, 184 bp), and (c) A/A homozygote (55 bp, 107 bp, 129 bp) [104].

temperature gradient gel electrophoresis (TGGE). The authors used the principle of single-strand conformational polymorphism (SSCP) for DNA mutation detection. A temporal thermal gradient was induced in the microfluidic network by controlling a heating block. The sequence variants in a heteroduplex sample of a 100 bp DNA fragment were distinguished. Schmalzing et al. [107] described a microfluidic device combined with EMD for clinical mutation studies. The authors reported the method was accurate without sample clean up. Russom et al. [108] reported allele specific extension of fluorescently labeled nucleotides for scoring of single nuleotide polymorphism (SNP) on a microchip. The authors also described the detection of single nucleotide incorporation by using a DNA pyrosequencing technique, based on the detection of pyrophosphate released during DNA synthesis. Liu et al. [109] reported the separation of the homoduplex and heteroduplex PCR products using chip-based temperature gradient electrophoresis. A 2-DE chip incorporated with the temperature gradient feature was used to analyze SNP in a single run. Zhang et al. [110] described temperature gradient nanocapillary electrophoresis (TG-NCE) for DNA mutation/single nucleotide polymorphism (SNP) analysis. According to the authors TG-NCE analyses of four mutant DNA samples, amplified from plasmid templates, showed that mutations were successfully detected under a wide temperature gradient of 10°C. The authors advocated the effectiveness of their system by demonstrating the successful detection of *K-ras* gene mutations in six colon cancer cell lines.

8.3.3. Amino Acids

Amino acids are the building blocks of proteins, the major constituent of our body. Almost all analytical methods have been tested for application to analysis of amino acids. Pumera [111] presented a review article on the analyses of amino acids using microfluidic devices. The author discussed the progress that has been made on the development of all steps needed for fully integrated microfluidic amino acids analyzer, sample introduction, preconcentration, self-calibration, derivatizations, and reactions. Harrison and coworkers [112] used 1 to 10 cm long capillaries etched on glass (10×30 μm) for separation of amino acids, with up to 75,000 theoretical plates in about 15 seconds. Zhang and Manz [113] described an NCE system and used it for analysis of FITC-labeled amino acids in both aqueous and binary media. The authors advocated this unit as an online monitoring unit due to its short residence time and small sample flow rate. The chip was made of glass with 1.5 μm thickness and a PDMS layer of 0.3 μm thickness. Furthermore, Zhang and Yin [114] reported NCE separation of FITC-labeled valine and alanine amino acids by using 20 mM sodium phosphate buffer.

Wu et al. [115] developed an integrated microfluidic device for analyzing the chemical contents of a single cell. The device comprised four different functions, that is, cell handling, metering and delivery of chemical reagents, cell lysis and chemical derivatization, and derivatized amino acid detection by LIF. Mourzina et al. [116] described NCE for the analysis of amino acids on a hybrid PDMS-glass chip. The separation was optimized by current-voltage linearity, contact angle, electroosmotic velocity, electroosmotic mobility, and electrokinetic potential. Abad-Villar et al. [117] described analyses of biochemical species on electrophoresis chips with an external contactless conductivity detector. The molecules studied were various amino acids. The detection was carried out on bare electrophoresis chips made from PMMA by probing the conductivity in the channel with a pair of external electrodes. Furthermore, the authors reported separation efficiencies up to 15,000 plates. Fu et al. [118] described analyses of sodium fluorescein and fluorescein isothiocyanate-labeled amino acids by using NCE-LIF with 1.1 pM as detection limit. Shen et al. [119] described a microfluidic array system with NCE for analyses of series biomolecules, including amino acids, proteins, and nucleic acids, with good reproducibility.

Sun et al. [120] developed PMMA chips and used them in electrophoresis for baseline separation of fluorescently labeled amino acid within 15 seconds. According to the authors theoretical plate numbers were 5000 for an approximately 3 cm separation distance. The authors claimed it as a new solvent imprinting and bonding approach, significantly simplifying the process for fabricating microfluidic structures in hard polymers such as PMMA. Shadpour et al. [121] described a 16-channel microfluidic chip with an integrated contact conductivity sensor array for analysis of amino acid. The electric field applied was 90 V/cm. Shackman et al. [122] fabricated a low cost, polymeric, eight-channel multiplexed microfluidic device to analyse amino acids and immunoassay products via gradient elution moving boundary electrophoresis (GEMBE). Ceriotti et al. [123] described an integrated fritless column in NCE with conventional stationary phases of octadecylsilanized silica gel. The analyses of FITC-labeled amino acids were completed within 15 seconds using this unit.

8.3.4. Drug Development and Design

In situ pharmacokinetics and pharmacodynamics are essential issues in drug development. After metabolism, the drug and its metabolites are found in blood and urine at extremely low concentrations. Sometimes the amount of biological fluids is very low especially in infants and the cerebrospinal fluid. Therefore, development of microdevices have been a boon for drug discovery programs. NCE has also achieved a great reputation in the drug development arena. Many authors have used this technique to analyze various drugs *in vivo* and *in vitro*. Liu et al. [124] developed a microfluidic NCE method based on indirect LIF detection to study protein-drug interactions. The authors studied interactions of heparin and BSA as a model system. These workers used sodium fluorescein as background and redistilled water as a marker to monitor EOF. Furthermore, the electrophoretic mobility changes of BSA were measured by taking various concentrations of heparin added to the running buffer. Each run was completed within 80 seconds, with binding constant of $1.24 \pm 0.05 \times 10^3 \, M^{-1}$. Hop [125] used silicon chip-based nanoelectrospray devices as more practical than pulled capillaries in NCE for determination of pharmacokinetics and metabolites.

Ramseier et al. [126] described analysis of FITC-derivatized amphetamine, methamphetamine, 3,4-methylenedioxymethamphetamine, and β-phenylethylamine in human urine by NCE. The authors advocated this unit as fast and highly efficient with no loss of accuracy and precision. Beard and de Mello [127] reported analyses of biogenic amines (putrescine, histamine, and tryptamine) on a PDMS-glass NCE with fluorescence detection (Fig. 8.15). The buffer used was phosphate and 2-propanol, and rhodamine 110 as a background fluorophore. According to the authors, this device provided sample preparation stages, an attractive technique for the analysis of biogenic amines. Furthermore the same group [128] described integrated on-chip derivatization and electrophoresis analysis of biogenic amines.

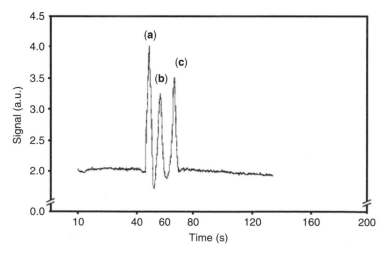

Figure 8.15 Electrophoretic separation of biogenic amines using a counterionic Indirect Fluorescence Detection (IFD) system, (a) putrescine, (b) histamine, and (c) tryptamine [127].

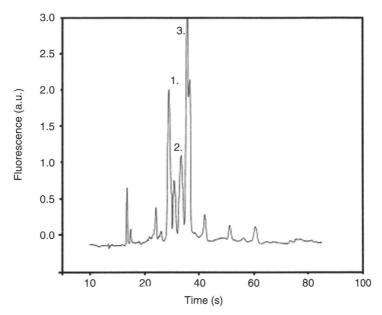

Figure 8.16 The separation of DTAF-biogenic amines in a Thai fish sauce sample. Peaks: (1) histamine, (2) putrescine, and (3) dichlorotriazine fluorescein (DTAF) [128].

In this work the authors presented the use of dichlorotriazine fluorescein (DTAF) as a superior *in situ* derivatizing agent for biogenic amines in microfluidic devices. The separation was achieved within 60 seconds, with a detection limit of 1.0 nM. The device was used to identify biogenic amines in Thai fish sauce sample (Fig. 8.16). Schwarz [129] described enzyme catalyzed amperometric oxidation of neurotransmitters (biogenic monoamines) in NCE. The detection limit achieved was 10^{-7}–10^{-8} M. Varjo et al. [130] reported the separation of FITC-labeled amines (ephedrine, 2,2-diphenylethylamine, 2-octylamine, and 2-butylamine) by NCE in uncoated and polyvinyl alcohol-coated glass chips using water and dimethyl sulfoxide (DMSO) as solvents of background electrolyte. The best separation of these amines was achieved on the coated channel. Furthermore, the separation was successful in nonaqueous DMSO electrolyte solution containing ammonium acetate and sodium methoxide, on both uncoated and coated glass microchips.

Deng et al. [131] described analysis of carnitine, acetylcarnitine, imipramine, and desipramine in human plasma using NCE-MS. The method developed may be useful for bioanalytical measurements for these important compounds in synthetic mixtures. Liu et al. [132] developed an NCE unit used for analysis of dopamine, with a detection limit of 100 nM. Kim et al. [133] described NCE based on a PDMS-glass chip for catechol and dopamine with electrochemical detection. The separation was achieved within 80 seconds by using a separate electric field of 60 V/cm. Johirul et al. [134] described NCE coupled with a cellulose-single stranded DNA (cellulose-ssDNA) modified electrode for the analyses of dopamine, norepinephrine, 3,4-dihydroxy-L-phenylalanine (L-DOPA), 3,4-dihydroxyphenylacetic acid (DOPAC), and ascorbic acid. These compounds were separated on a 62 mm long separation channel at the separation field strength of 200 V/cm within 220 seconds in a 10 mM phosphate buffer (pH 7.4). The most favorable potential for the amperometric detection was 0.7 V (vs. Ag/AgCl).

Schulze et al. [135] developed fused-silica chips dynamically coated with hydroxypropylmethyl cellulose and utilized them for the separation of aromatic low molecular weight compounds such as serotonin, propranolol, a diol, and tryptophan. The authors used deep UV laser-induced fluorescence detection for these compounds. Schuchert-Shi et al. [136] identified ethanol, glucose, ethyl acetate, and ethyl butyrate, byproducts obtained in enzymatic conversions using hexokinase, glucose oxidase, alcohol dehydrogenase, and esterase. The authors reported that the quantification for ethyl acetate was possible using contactless conductivity detection. Hu et al. [137] described the separation of reaction products of β-thalassemia in a multiplex primer-extension reaction using NCE. The method developed was used for patient samples and the results coincided with those of a detection kit.

8.3.5. Enzymes and Hormones

Enzymes and hormones are found in very low concentrations in the cells and blood of various animals, and their separation and identification by conventional analytical techniques are quite tedious. NCE has been used to solve this problem, and some workers have attempted to analyze these identities using this technique. Murakami et al. [138] described NCE for alkaline phosphatase enzyme testing. Wang et al. [139] reported analysis of glucose in connection to the corresponding glucose oxidase (GOx) and glucose dehydrogenase (GDH) reactions by using NCE. The authors used peak current ratio for confirming the peak identity, estimating the peak purity, addressing co-migrating interferences and deviations from linearity. A voltage of 2000 V/cm resulted in peroxide and NADH migration times of 93 and 260 seconds, respectively. Furthermore, the same group [140] used NCE for analyses of multienzymatic dehydrogenase/oxidase. The operation of oxidase/dehydrogenase reaction/separation in NCE was illustrated for the simultaneous measurement of glucose and ethanol in connection to the corresponding glucose oxidase and alcohol dehydrogenase reactions, respectively. A voltage of 2000 V/cm resulted in peroxide and NADH migration times of

74 and 230 seconds, respectively. The optimized parameters were reaction conditions and separation experimentation.

Starkey et al. [141] reported the determination of endogenous extracellular signal-regulated protein kinase by NCE. It has been reported that the microchip assay provided a rapid and accurate alternative to conventional methods. Zhuang et al. [142] described a chip-based electrophoresis and on-column enzymatic reaction analysis protocol for lactate dehydrogenase (LDH) isoenzymes with a homemade xenon lamp-induced fluorescence detection system. The authors used a four-step operation and temperature control for the determination of LDH activity on-chip monitoring of incubation product of NADH during the fixed incubation period and at a fixed temperature. The authors carried out experiments on the determination of LDH standard sample and serum LDH isoenzymes from a healthy adult donor. Dishinger and Kennedy [143] used NCE to monitor the secretion of insulin from islets of Langerhans. The device was used to complete over 1450 immunoassays of biological samples in less than 40 min, allowing the parallel monitoring of insulin release from four islets every 6.25 seconds with 10 nM as detection limit. Roper et al. [144] also described an NCE system for determination of insulin secreted from islets of Langerhans. The separation was completed within 5 seconds by using an electric field of 500 V/cm with 3.0 nM as detection limit. Xu et al. [145] presented interfacing capillary gel microfluidic chips with infrared laser desorption mass spectrometry for analysis of bovine insulin and bradykinin. Henares et al. [146] used NCE for analyses of proteases (trypsin, chymotrypsin, thrombin, elastase) by applying the drop-and-sip technique.

Licklider et al. [147] described an NCE-MS unit on a silicon chip for analysis of proteins obtained from trypsin digest of cytochrome c (Fig. 8.17). The authors described their work as an efficient electrospray interface integrated with mass spectrometry. Similarly, Wu and Chen [148] designed a sheathless NCE/ESI-MS using an electrodeless nanospray interface, with a pulled bare capillary tip as an ESI emitter. The tip was fabricated by applying a small weight on the lower end of a vertical capillary section. The separation and identification of cytochrome c tryptic digest by online NCE/ESI-MS is shown in Fig. 8.18 indicating good resolution. The experimental conditions were acetonitrile-0.5% acetic acid (1 : 1, v/v) buffer with flow rate of 52.5 nL/min at 10 kV/cm applied voltage. Ling et al. [149] described a microchip electrophoresis method for simultaneous determination of reactive oxygen species (ROS) and reduced glutathione (GSH) in the individual erythrocyte cell. The authors used cell sampling, single-cell loading, docking, lysing, and capillary electrophoretic separation with LIF detection integrated on a microfluidic chip with crossed channels. The authors labeled ROS with dihydrorhodamine 123 in the intact cell and GSH with 2,3-naphthalene-dicarboxaldehyde, which were

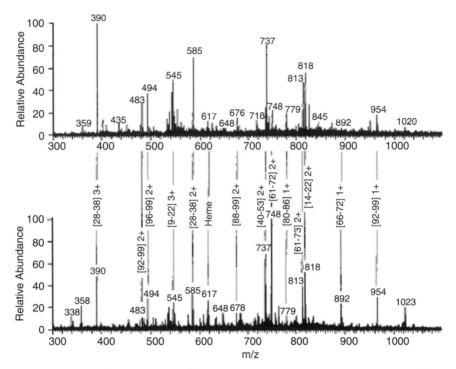

Figure 8.17 Comparison of MS/MS analysis of a peptides mixture obtained from a trypsin digest of cytochrome c using a pulled, borosilicate, nanospray emitter (lower spectrum) and a micromachined parylene emitter (upper spectrum) [147].

included in the separation medium. The microfluidic network was optimized to prevent cell leaking from the sample reservoirs into separation during the separation phase. The detection limits of 0.5 and 6.9 amol for ROS and GSH, respectively, were achieved. Qin et al. [150] reported NCE-LIF for simultaneous determination of two kinds of intracellular signaling molecules (ROS and GSH) related to apoptosis and oxidative stress. As the probe dihydrorhodamine 123 (DHR-123) was converted intracellularly by ROS to the fluorescent rhodamine 123 (Rh-123), and the probe naphthalene-2,3-dicarboxaldehyde (NDA) reacted quickly with GSH for producing a fluorescent adduct, rapid determination of Rh-123 and GSH was achieved on a glass microchip within 27 seconds by using a 20 mM borate buffer (pH 9.2). This method was tested to measure the intracellular ROS and GSH levels in acute promyelocytic leukemia (APL)-derived NB4 cells. Typical electropherograms of standard solutions of GSH (1.061025 M) derivatized with NDA (1.061024 M) and Rh-123 (6.5610211 M) are shown in Fig. 8.19. The authors studied the

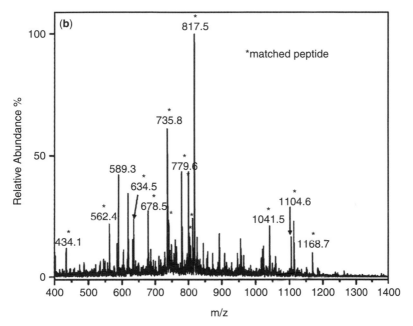

Figure 8.18 ESI mass spectrum of cytochrome c (43.8 fmol) tryptic digest obtained by summing over the range 18.5 to 19.3 min [148].

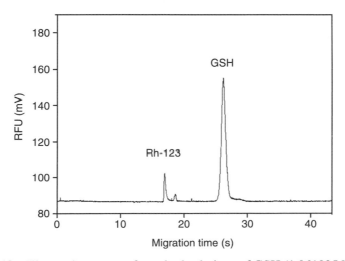

Figure 8.19 Electropherogram of standard solutions of GSH (1.061025 M) derivatized with NDA (1.061024 M) and Rh-123 (6.5610211 M) [150].

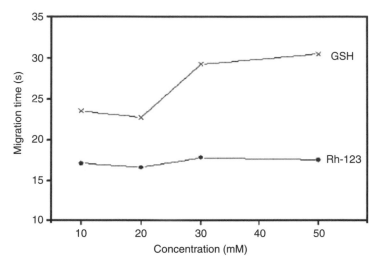

Figure 8.20 Effect of buffer concentration on the migration times of Rh-123 and GSH. Separation conditions: injection time 12 seconds, separation voltage 1300 V/cm, and effective separation distance 10 mm [150].

effects of influence of buffer concentration and applied voltage on the migration times of Rh-123 and GSH; these are shown in Fig. 8.20 and Fig. 8.21, respectively, indicating high migration times at high buffer concentration and low applied voltage. Furthermore, the effect of pH on the fluorescence intensity of Rh-123 and GSH is shown in Fig. 8.22 indicating a low limit of detection of both species at around 8.0 pH.

8.3.6. Biological Fluids

Microfluidic devices have gained importance and utility for analyses of various molecules, including drugs and their metabolites. Vrouwe et al. [151] developed NCE for point-of-care testing of lithium in blood samples. The device consisted of a glass chip coupled with a conductivity detector. The authors tested this system for lithium analysis in five patients in the hospital. Furthermore, the authors reported that sodium, lithium, magnesium, and calcium were separated in <20 seconds. The authors claimed that the NCE system provided a convenient and rapid method for point-of-care testing of electrolytes in serum and whole blood.

Zhuang et al. [152] derivatized glycan samples with 8-aminopyrene-1,3,6-trisulfonic acid to get a charge for electrophoresis and a fluorescent label for detection. They analyzed glycan derivatives in the blood of cancer patients

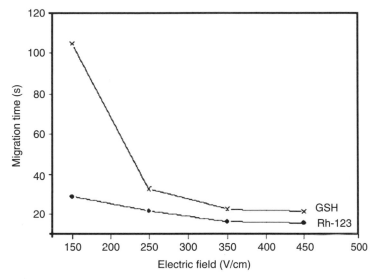

Figure 8.21 Effect of applied voltage on the migration times of Rh-123 and GSH in 20 mM borate buffer (pH 9.2). Separation conditions as in Fig. 8.20 [150].

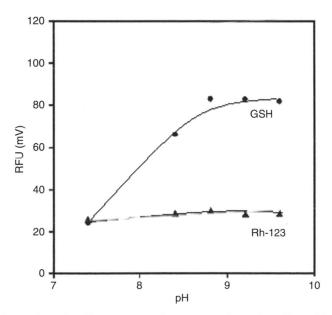

Figure 8.22 Effect of buffer pH on the fluorescence intensity of Rh-123 and GSH in 30 mM borate buffer. Separation conditions: injection time 12 seconds, separation voltage 300 V/cm, and effective separation distance 10 mm [150].

using NCE. Maeda et al. [153] described microchip electrophoresis on a plastic chip for analysis of glucose in human blood samples. According to the authors treatment of plasma with hexokinase or glucokinase for glucose phosphorylation resulted in a peak shift from 145 to 70 seconds, corresponding to glucose and glucose-6-phosphate, respectively. The reproducibility of the assay was found to be about 6.3% to 9.1% (RSD). The within-days and between-days reproducibility were 1.6% to 8.4% and 5.2% to 7.2%, respectively. Phillips [154] described an NCE system for analyzing the concentrations of inflammatory cytokines in the cerebrospinal fluid of patients with head trauma. The separation of cytokines was achieved within 2 minutes with quantification of the resolved peaks being achieved by online LIF and integration of each peak area. Gao et al. [155] described an integration of single cell injection, cell lysis, separation, and detection of intracellular constituents in NCE with an LIF detector. The channel size was 12 μm deep, 48 μm wide, and 35 mm long. The cell lysis occurred within 40 ms under an applied voltage of 1.4 kV/cm. Vrouwe et al. [156] described analysis of lithium in blood samples. Du and coworkers [157] reported glucose in human plasma by using NCE coupled with electrochemical detection.

Deng et al. [158] developed NCE-MS for analysis of carnitines in human urine within 48 seconds. The results demonstrated the feasibility for NCE separation and electrospray mass spectrometric detection for these important compounds in synthetic mixtures, as well as in human urine extracts. Wang et al. [159] used NCE for the determination of glucose, uric acid, ascorbic acid, and acetaminophen. The fluid control was used to mix the sample and glucose oxidase enzyme (GOx) and the enzymatic reaction, a catalyzed aerobic oxidation of glucose to gluconic acid and hydrogen peroxide, occurred along the separation channel. Chan and Herold [160] used an automated NCE with fluorescent detection to quantify total microalbuminuria, an important prognostic marker in diabetic nephropathy and cardiovascular disease. The authors reported that NCE could detect both immunoreactive and nonimmunoreactive forms of albumin. This system was a simple, robust method to quantify microalbuminuria, with good sensitivity and precision. Wilke and Buttgenbach [161] described the analyses of hydrogen peroxide, ascorbic acid, and uric acid using NCE with amperometric detection. All three oxidizable species were detected in less than 70 seconds. The electropherograms are shown in Fig. 8.23 and Fig. 8.24 indicating good separation of these species. Žuborova et al. [162] measured oxalate in urine using NCE with conductivity detection. The buffer used was 15 mM propionic acid, ε-aminocaproic acid (pH 4.0) with 0.05% methylhydroxyethyl cellulose. Similarly, Fanguy et al. [163] measured uric acid in human urine

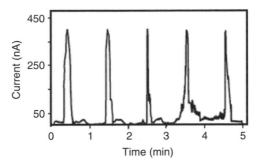

Figure 8.23 Electropherogram for multiple injections of 10 mM hydrogen peroxide with phosphate buffer (10 mM, pH 7.4), separation potential 400 V/cm, injection potential, 400 V/cm and detection potential 700 mV/cm [161].

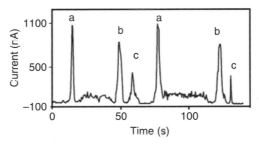

Figure 8.24 Electropherogram for sample solution containing 50 mM hydrogen peroxide. (a) Hydrogen peroxide, (b) ascorbic acid, and (c) uric acid. Experimental conditions as in Fig. 8.23 [161].

by NCE with amperometric detection. The experiments were performed in 25 mM/L MES (pH 5.5) and completed within 30 seconds.

8.3.7. Foods and Beverages

NCE has played a crucial role in the foods and beverages industries by identifying various compounds at low levels. Suzuki et al. [164] reported analysis of amino sugars using NCE-LIF (33 mm long separation channel) within 60 seconds, with a detection limit of 0.5 fmol. The separated molecules were D-glucosamine and D-galactosamine and their reduced forms were labeled with 4-nitro-2,1,3-benzoxadiazole 7-fluoride (NBD-F). The method was applied for analysis of amino sugar compositions of O-linked glycans released from bovine submaxillary mucin with alkali in the presence of borohydride. Usui et al. [165] reported a polymeric 2-DE chip system and demonstrated its applications for commercially available proteins as a standard specimen and tissue-extracted proteins as the real samples. The authors advocated this unit as an improved labor-intensive operation with reduced experimental time.

Zhang et al. [166] used NCE for the separation of monosulfate glycosaminoglycan disaccharide isomers by addition of 1,4-dioxane (DO) dramatically in BGE. Furthermore, the authors used methylcellulose to suppress EOF and analyte adsorption to the chip. To improve analyte resolution, buffer pH, β-CD, and DO were investigated and fast separation was achieved by increasing electric field strength and field-amplified sample stacking with increasing buffer concentrations. The authors used this method for separation and identification of monosulfate and trisulfate unsaturated disaccharides (DeltaDi-UA2S, DeltaDi-4S, DeltaDi-6S, and DeltaDi-triS) derivatized with 2-aminoacridone hydrochloride. A mixture of monosulfate disaccharide isomers (DeltaDi-UA2S, DeltaDi-4S, and DeltaDi-6S) was baseline separated within 75 seconds on a PMMA chip by using a mixed buffer (DO/running buffer 57 : 43 v : v), 0.5% MC, pH 6.81, with an E(sep) of 558 V/cm. The theoretical plates were in the range of 5×10^5 to 1×10^6 per meter.

Masár et al. [167] reported quantitation of sulfate on a chip after an in-sample oxidation of total sulfite in wine. The authors reported 99% to 100% recoveries of sulfite, determined for appropriately spiked wine samples, indicating a very good accuracy of the method. Masár et al. [168] described analysis of organic acids in wine by NCE. Approximately 22 organic and inorganic acids in wines were analyzed on a PMMA chip with integrated conductivity detection. Skelley and Mathies [169] described NCE analyses of α- and β-D-glucosamine and their interconversion in solution. pH-dependent mutarotation of α and β isomers were also studied. The mutarotation monitored by this device is shown in Fig. 8.25, indicating no change in glucosamine-1 peak at high pH while the peak of glucosamine-2 moves to longer migration times. The interconversion rate constants were reported to be 0.72 ± 0.09, 1.3 ± 0.1, and $2.2 \pm 0.3 \times 10^{-3}$ per second at pH 8.99, 9.51, and 10.01, respectively, by determining plateau and peak heights relative to the migration times. The effect of time on mutarotation is also shown in Fig. 8.26 indicating maximum mutarotation at 2609 seconds initial run time. Furthermore, the separation of a mixture of glucosamine, amines, diamines, and amino acids at pH 9.78 was reported, as shown in Fig. 8.27, indicating fast separation with sharp peaks. Methyl and ethylamine were well resolved from glucosamine anomer peaks even though they migrated between two species. Jayarajah et al. [170] used portable NCE for analyses of neurologically active biogenic amines (tyramine and histamine) in fermented beverages. According to these workers, the concentrations of tyramine and histamine were 1.0 to 3.4 and 1.8 to 19 mg/L, respectively.

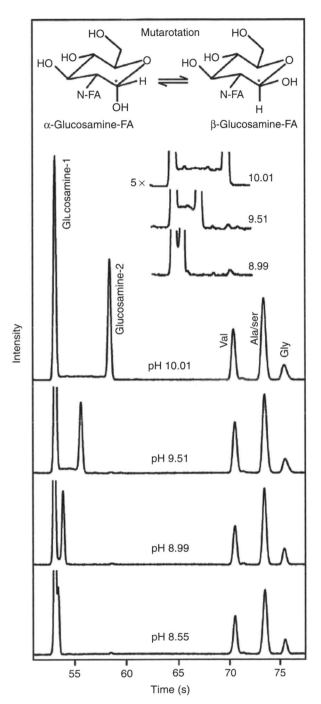

Figure 8.25 Electropherograms of glucosamine isomers as a function of pH [169].

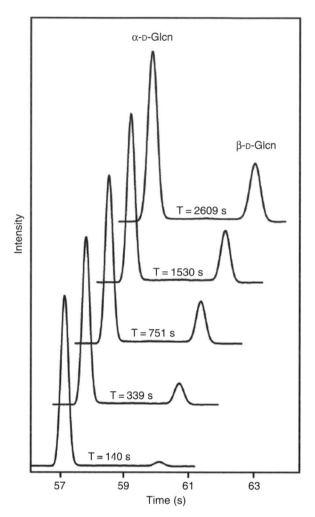

Figure 8.26 Analysis of α-D-glucosamine mutarotation as a function of time at pH 9.51 [169].

8.3.8. Viruses and Bacteria

Viruses and bacteria are the smallest biologically active identities discovered so far and, naturally, require nanodevices for their profiling. Analysis of viruses and bacteria on microchip devices are of particular interest for experiments with infectious material because of easy containment and disposal of samples. Thus, the use of NCE in routine analyses of viruses and bacteria is exceptionally attractive [171,172]. These pathogens are detected based on their different migration times without the need for time consuming sample preparation [173,174].

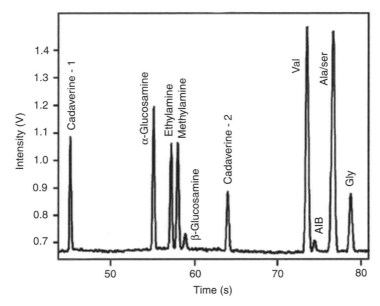

Figure 8.27   Electropherogram of glucosamine and a mixture of amines, diamines, and amino acids at pH 9.78 [169].

Zhou et al. [175] described the determination of severe acute respiratory syndrome (SARS) coronavirus by a microfluidic chip system. The unit included an LIF microfluidic chip analyzer, a glass microchip for both PCR and capillary electrophoresis, a chip thermal cycler based on dual Peltier thermoelectric elements, a reverse transcription-polymerase chain reaction (RT-PCR) SARS diagnostic kit, and a DNA electrophoretic sizing kit. According to the authors, the system allowed efficient DNA amplification of the SARS coronavirus followed by electrophoretic sizing and detection on the same chip.

Huang et al. [176] described an integrated microfluidic chip (of PDMS and soda-lime glass) capable of performing DNA/RNA amplification, electrokinetic sample injection and separation, and online optical detection in an automatic mode. The authors tested its functionality for bacterial DNA of *Streptococcus pneumoniae* and RNA of dengue-2 virus. The NCE developed represented a crucial contribution to the fields of molecular biology, genetic analysis, infectious disease detection, and other biomedical applications.

Tsai and Sue [177] designed a micro-RT-PCR chip for quantitative detection of tumor viruses. The authors integrated test sample reservoirs, RT-PCR meanders, and electrophoresis on an SU-8-based monolithic chip. Kolivoška et al. [178] reported the analysis of human rhinovirus serotype 2 (HRV2) on a commercially available NCE via labeling the proteinaceous capsid of HRV2 with Cy5 for detection by a red laser (λex 630 nm). The resolution of the sample constituents (virions, a contaminant present in all virus preparations, and excess dye) was improved upon adaptation of the separation conditions, mainly by adjusting the SDS concentration of the BGE. Furthermore, the same authors [179] reported NCE coupled with LIF detection. The authors used fluorescent dye for labeling viral capsids and analyzed HAV2 and subviral particles, followed by the complexation of the virus with a synthetic fragment of the VLDL-receptor. Lagally et al. [180] reported the separation and differentiation between strains of bacteria within 10 min, with detection limit of two bacterial cells in a 200 nL reactor on a microchip. Lantz et al. [181] reported detection limits down to a single cell under controlled sample conditions. The viability of applying NCE to cells or spores depends on a number of factors, including the extent of heterogeneity of the sample, cell viability within the sample matrix, and the extent of the differences in electrophoretic mobility among the different biological entities.

8.3.9. Explosives

In the few decades the activities of terrorists have increased all over the world and they are using a variety of weapons comprising different explosives. Normally, the amounts of explosives in the atmosphere after the blast are low, which requires microfluidic devices such as NCE in forensic science. The determination of explosive in the environment due to military activities is also a challenging area. Some workers have attempted to develop NCE methods for analyses of various explosives. Pumera [182] presented a review article on analyses of explosives using NCE. The most important explosives are (1) nitrated organic compounds, such as 2,4,6-trinitrotoluene (TNT), hexahydro-1,3,5-trinitro-1,3,5-triazine (RDX), methyl-2,4,6-trinitrophenylnitramine (tetryl), and nitroglycerin; (2) inorganic nitrate, chlorate, or perchlorate salts (NH_4NO_3, KNO_3; or NH_4ClO_4); and (3) compounds containing an unstable peroxide group (triacetone triperoxide). Nitrated organic explosives have been used for more than 100 years but, nowadays, RDX and HMX (octahydro-1,3,5,7-tetranitro-1,3,5,7-tetrazocine) are very powerful explosives widely used in various activities.

Wang et al. [183] reported fast analysis of preblast and postblast cations and anions by using NCE in 60 seconds. Low EOF of PMMA chip material facilitated the rapid switching between analyses of cations and anions using the same microchannel and run buffer, and provided rapid measurement of seven explosives-related cations and anions. Wang et al. [184] discussed the future development of an NCE device for the detection of triacetone triperoxide. The advantage, from UV degradation of triacetone triperoxide to hydrogen peroxide and from an already developed microfluidic-amperometry method for rapid detection of organic peroxides, was presented. Wang and Pumera [185] described a dual electrochemical microchip detection system containing two orthogonal detection modes (conductivity and amperometry) that facilitated the measurements of inorganic explosives and nitroaromatic explosive components on a single-channel microchip. The total assay of explosive mixture took 2 minutes, as shown in Fig. 8.28. Bromberg and Mathies [186] described a homogeneous immunoassay for detection of TNT and its analogues on a microfabricated capillary electrophoresis chip. The assay was based on the rapid electrophoretic separation of an equilibrated mixture of an anti-TNT antibody, fluorescein-labeled TNT, and unlabeled TNT or its analogue, with 1.0 ng/mL as the limit of detection. Furthermore, Bromberg and Mathies [187] described a high throughput homogencous

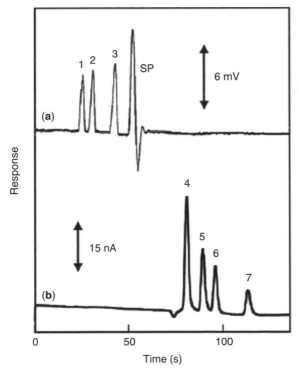

Figure 8.28 Electropherograms of inorganic and nitroaromatic explosives. (a) Amperometric and (b) conductivity detectors. Analytes: (1) ammonium, (2) methylammonium, (3) sodium, (4) TNB, (5) TNT, (6) 2,4-DNB, and (7) 2-Am-4,6-DNB [185].

immunoassay for the detection of TNT using NCE. The limit of detection reported was 1.0 ng/mL.

8.3.10. Environmental Analyses

Thousands of pollutants contaminate our environment, including organic, inorganic, and biological species. It has been reported that, generally, contaminants are found in the environment at trace levels [188,189]. Of course, conventional analytical methods have been used for their monitoring but these methods cannot detect contaminants present at extremely low concentrations. This challenge was overcome by newly developed microfluidic devices such as NCE.

Wang and coworkers [24] described NCE with a thick film amperometric detector for separation and detection of organophosphate nerve agents (paraoxon, methyl parathion, fenitrothion, and ethyl parathion) within 140 seconds. The electropherograms are shown in Fig. 8.29 indicating good separation. The separation conditions were buffer, 20 mM MES (pH 5.0) containing 7.5 mM

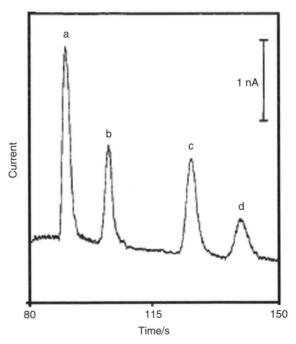

Figure 8.29 Separation and detection of organophosphate nerve agent compounds: (a) 1.0×10^{-5} M paraoxon, (b) 1.0×10^{-5} M methyl parathion, (c) 2.0×10^{-5} M fenitrothion, and (d) 4.0×10^{-5} M ethyl parathion [24].

SDS, separation voltage 2000 V/cm, injection voltage 1500 V/cm, injection time 3 seconds, and detection potential -0.5 V (vs. Ag/AgCl). Furthermore, the same group [190] described NCE-based separation of organophosphonate nerve agent (sarin and soman) degradation products (alkyl methylphosphonic acids) with contactless conductivity detection. The system has been used for monitoring these agents in river water sample within 50 seconds.

Lu et al. [191] described determination of cyanide in vapor and liquid phases in NCE within 40 seconds. LIF detection was used to monitor the fluorescent isoindole derivative formed by the reaction of cyanide with 2,3-naphthalenedicarboxaldehyde (NDA) and taurine. Li et al. [192] described speciation of arsenic species using online NCE hyphenation with hydride generation atomic fluorescence spectrometry (HG-AFS). A baseline speciation of arsenite and arsenate within 54 seconds, on a 90 mm long channel having a chip at 2500 V/cm potential with a mixture of 25 mM H_3BO_3 and 0.4 mM CTAB (pH 8.9) as electrolyte buffer, was achieved. According to the authors the precision (RSD, $n = 5$) ranged from 1.9% to 1.4% for migration time of 2.1% to 2.7% for peak area, and 1.8% to 2.3% for peak height for the two arsenic species. The electropherograms of arsenic speciation are shown in Fig. 8.30 indicating a good speciation within 55 seconds.

Li and coworkers [193] reported a hybrid technique for rapid speciation analysis of Hg(I) and MeHg(II) by directly interfacing an NCE to atomic fluorescence spectrometry. Both mercury species were separated as their cysteine complexes within 64 seconds. The precision (RSD, $n = 5$) of migration time,

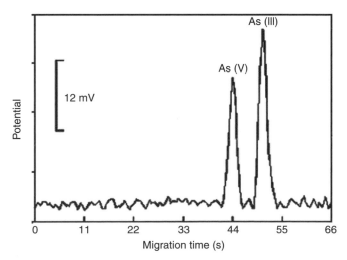

Figure 8.30 Electropherogram of arsenite and arsenate species of 3.0 mg/L concentrations [192].

peak area, and peak height for 2 mg/L Hg(II) and 4 mg/L MeHg(I) (as Hg) ranged from 0.7% to 0.9%, 2.1% to 2.9%, and 1.5% to 1.8%, respectively. Henares et al. [146] described an NCE integrated with the plural different reagent-release capillaries, acting as various biochemical sensors. The authors reported a novel drop-and-sip technique of fluid handling for analyses of divalent cations [Ca(II), Zn(II), and Mg(II)]. Vogt et al. [194] described a new two-chip concept in terms of stability and reproducibility of separation and detection of sodium, potassium, and lithium. The RSD of <1% and 3%, respectively, for retention times and peak heights during long-term measurements could be achieved. Hui et al. [195] described an interface of NCE-ICP on a PDMS chip. A stainless steel tube was placed orthogonal to the exit of the NCE separation channel for cross flow nebulization with a supplementary flow of buffer solution at the channel exit. Two capillaries were inserted into the NCE chip near the inlet of the separation channel for sample and buffer solution injection. The nebulization and transport efficiency of the NCE-ICP interface was approximately 10%. Barium and magnesium ions were eluted from the NCE within 30 seconds, applying 1 kV/cm as applied potential, with good resolution, as shown in Fig. 8.31. Kutter et al. [196] analyzed magnesium and calcium by NCE after sample stacking and on-chip complexation with 8-hydroxyquinolin-5-sulfonic acid.

Rohlícek and Deyl [197] described a versatile tool allowing standard operations (i.e. washing, preconditioning, separation, and inner surface modification

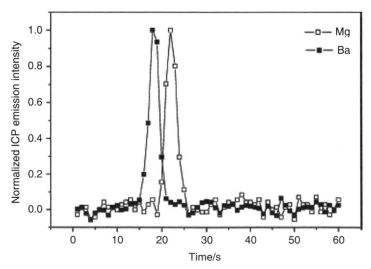

Figure 8.31   Electropherograms of barium and magnesium ions using NCE [195]. Potential was 1 kV. −□− Mg emission; −■− Ba emission.

of the chip channel) with capillary electrophoresis chips. Though currently designed for a chip of maximal dimensions (30 × 60 mm), other formats of the chip required only a minimum adjustment of the equipment, namely setting of the chip sliding rails and adequate arrangement of the exchangeable heads. The applications of the tool are demonstrated by the separation of the standard set of inorganic cations. Wang et al. [198] described NCE for analysis of phenolic compounds. Furthermore, the effects of the sample flow rate, applied voltages, and other relevant variables were described and it was found that the peak intensity was independent of the flow rate. Le Saux et al. [199] described the binding constant of 2-naphthalenesulfonate and neutral phenols (phenol, 4-chlorophenol, and 4-nitrophenol) into β-cyclodextrin by using NCE. The authors discussed the issue of competitor choice in relation to its appropriateness for proper monitoring of the interaction.

Fleger et al. [200] described a coupling between multimode polymer waveguides and microfluidic channels on a PMMA chip with optical detection for the dye sulfanilazochromotrop (SPADNS). Edel et al. [201] reported a thin film polymer light emitting diode as integrated excitation source in NCE for the separation of fluorescent dyes. Feng et al. [202] described a portable NCE system with potential gradient detection (PGD) for analyses of alkali metals and alkaloids. The design of the system showed several advantages, such as simplicity, miniaturization, and wide applicability. Nikcevic et al. [203] described molded PMMA (IM-PMMA) chips as potential candidates for electrophoresis. The authors evaluated several important properties of IM-PMMA chips, such as fabrication quality, scanning electron microscope imaging, surface quality measurements, selected thermal/electrical properties, and the influence of channel surface treatments. The EOF was also evaluated for untreated and oxygen reactive ion etching (RIE)-treated surface microchips. The performance of single lane plastic microchip electrophoresis separations was evaluated by using a mixture of two fluorescein (FL) and fluorescein isothiocyanate (FITC) dyes. The reproducibility of average migration time ratios of FL and FITC for all chips is shown in Fig. 8.32.

Ujiie et al. [204] fabricated quartz chips for NCE and reported the separation of rhodamine B and sulforhodamine at 14.4 and 66.6 cm separator lengths. The buffer was 20 mM phosphate buffer at 2 kV applied voltage and the separation was achieved in 70 seconds. Wakida et al. [205] reported a high throughput characterization for dissolved organic carbon in environmental waters within 2 minutes using NCE. The authors collected water samples from 10 sampling points at the Hino River that flows into Lake Biwa. Shin et al. [206] described NCE (PDMS) with fluorescence detection for analyses of atrazine.

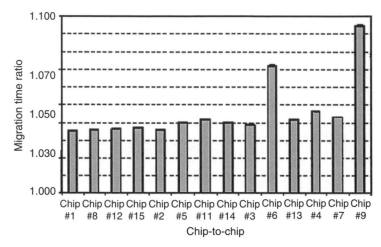

Figure 8.32   Representation of reproducibility of average migration times ratio of (fluorescein) FL and fluorescein isothiocyanate (FITC) for all chips [203].

8.3.11. Miscellaneous Applications

NCE has also been used to determine various bioprocesses, generally performed by measuring fundamental physical and chemical parameters, such as temperature, pressure, pH, and reaction rates. NCE has also been used to study some chemical reactions on chip followed by the separation of reaction products.

Brivio et al. [207] devised an NCE-NESI-MS system used for online monitoring of supramolecular interactions of Zn-porphyrin with pyridine, 4-ethylpyridine, 4-phenylpyridine, N-methylimidazole, and N-butylimidazole in acetonitrile, as well as host-guest complexations of β-cyclodextrin with N-(1-adamantyl)acetamide or 4-tert-butylacetanilide. Stettler and Schwarz [208] described a microchip electrophoresis instrument equipped with a UV-detector and a home built chip station with electrochemical detection for the calculation of interactions and to binding constants of ligands with their substrates (β-cyclodextrin substrate with neurotransmitters). Fuchs et al. [209] described a technology based on a microelectronic chip integrating an array of over 100,000 independent electrodes and sensors which allowed individual and parallel single cell manipulation up to 10,000 cells while maintaining viability and proliferation capabilities.

Henares et al. [210] developed a multiple enzyme linked immunosorbent assay (ELISA) chip by using capillary-assembled microchip (CAs-CHIP), which involved simple embedding of a 2 to 3 mm length of square capillaries possessing valving and immunoreaction functions into the microchannels fabricated on a PDMS substrate. The authors immobilized different anti-IgGs (human, goat, chicken) and used as ELISA parts of CAs-CHIP. According to the authors the valve integrated multi-ELISA chip developed could be applied for many different diagnostic purposes by using different immunoreaction capillaries necessary for specific clinical diagnostic applications.

Cheng et al. [211] described an NCE-LIF system for the rapid separation and quantitative detection of ligase chain reaction (LCR) products amplified from the *lac*I gene on a silicon-glass chip. LCR is a useful molecular technique for

TABLE 8.1 The Applications of NCE for Separation and Identification of Different Molecules

Compounds	Detectors	Chip Materials	LOD	Refs.
Mg(II) & Ca(II)	LIF	Glass	20–40 µM	[197]
Cr(III), Co(II) & Cu(II)	LIF	PDMS	493 nM	[213]
Co(II) & Cu(II)	LIF	Fused silica	12.5 nM	[214]
TNT, TNB & Tetryl	UV-Visb.	Glass	ppb–ppt	[215]
Biotin	MS	Silicon	100 fg	[216]
PSA	Optical reflection	Silicon nitride	0.2 ng	[217]
DNA polymerase	Diffraction	Silicon	4.7 ng	[218]
Proteins	Piezoresistance	Silicon nitride	10 ng	[219]
GCN$_4$ antigenic peptide	Optical reflection	Silicon	20 ng	[220]
DNA	Optical reflection	Silicon	10–100 µM	[221–223]
DNA	Piezoresistance	Silicon	75 nM	[224]
Butyl rhodamine B	Chemiluminescence	Glass	10^{-9} M	[225]
α-Lactalbumin, β-lactoglobulin A & β-lactoglobulin B	Fluorescent	Glass	<0.5 pg	[50]
Peptides	TOF-MS	Silica	3.2–43.5 nM	[52]
DNA	LIF	Glass	2 fM	[79]
DNA	–		1.83 ng/µL	[82]
DNA	–	PDMS	1 fmol	[97]
Amino acids	LIF	–	1.1 pM	[118]
Dopamine	ECD	PDMS	100 nM	[132]
Insulin	Fluorescent	Glass	10 nM	[143]
Insulin	Fluorescent	Glass	3.0 nM	[144]
Amino sugar	LIF		0.5 fmol	[164]
TNT	Fluorescent	Glass	1.0 ng/mL	[186]
TNT			1.0 ng/mL	[187]

Abbreviations: DNA, deoxyribose nucleic acid; ECD, electrochemical detection; LIF, laser induced fluorescence; LOD, limit of detection; PDMS, poly(dimethylsiloxane); PSA, prostate specific antigen; TOF-MS, time-of-flight mass spectrometer; TNT, 2,4,6-trinitrotoluene.

detecting known point mutations. The technology is used in disease diagnosis and research. Liao et al. [212] described PMMA-based NCE with electrochemical detection for in-column urea/urease reactions. The authors advocated this unit as capable of high resolution NCE-electrochemical detection of bioanalytical reactions. Some other applications of NCE are summarized in Table 8.1.

8.4. MECHANISM OF SEPARATION

As in the case of normal CE, small cations have higher charge/densities (q_i/r_i) ratio and, hence, larger ionic mobilities. As a principally striking electrophoretic property, it should lead to a rapid separation of high efficiency. But in practice, this seldom occurs unless special precautions are taken. The different charges on various species may be an advantage of their separation in NCE as the migration of these is controlled by the charges. In NCE sample injection, capillary and detection are integrated on a single chip with anode and cathode electrodes at the inlet (sample loading) and the detector sides, respectively. Therefore, analytes having positive charge move from anode to cathode where they are detected by a suitable detector. The movements of ionic species from anode to cathode are due to the electrophoretic (μ_{ep}) and electroosmotic (μ_{eo}) mobilities. Therefore, the migration of species is controlled by the sum of the electrophoretic and electroosmotic mobilities. The electrophoretic mobilites of the species depend on their charges and the sizes while there is no relation among the charges, sizes, and the electroosmotic mobility. The analytes of different charge/radii ratios migrates by different electrophoretic velocities under the influence of applied voltage. The greater the charge/radius ratio the greater is the mobility and, hence, the lower is the migration time. Therefore, analytes having greater charge always eluted first followed by species with smaller charge. The electroosmotic mobility for all analytes remains almost the same but it helps in their migration towards the cathode. In this way, species elute at different migration times due to the combined effect of different electrophoretic and electroosmotic mobilities. In addition, the interaction of analytes with the capillary wall, steric effect, van der Waal forces, dispersion interaction, etc., also play a crucial role for the different mobilities of the species.

8.5. CONCLUSION

Remarkable developments have been achieved in NCE and its applications for a wide range of analytes. The utility of NCE for exploring the profiles of pathogens is of great concern for the welfare of our society. NCE is still not fully developed. Of course, we believe that NCE will be a powerful tool in nanoanalyses. Moreover, it will reveal many mysteries of biological and environmental sciences due to its inherent character of nano or low level characterization. NCE is considered a big achievement in separation science and it will prove itself as a mature reliable analytical technique.

REFERENCES

1. F. Kohlrausch, *Ann. Phys.*, **62**, 209 (1897).
2. A. Tiselius, The moving boundary method of studying the electrophoresis of proteins, Ph.D. Thesis, *Nova acta regiae societatis scientiarum*, Ser. IV, Vol. 17, No. 4, Uppsala, Sweden: Almqvist & Wiksell (1930), 1–107.
3. S. Hjerten, *Chromatogr. Rev.*, **9**, 122 (1967).
4. J.W. Jorgenson, K.D. Lukacs, *Anal. Chem.*, **53**, 1298 (1981).
5. J.W. Jorgenson, K.D. Lukacs, *J. Chromatogr.*, **218**, 209 (1981).
6. J.W. Jorgenson, K.D. Lukacs, *Science*, **222**, 266 (1983).
7. T. Wehlr, M. Zhu, Capillary electrophoresis: Historical perspectives, in J.P. Landers, ed., *Handbook of capillary electrophoresis*, Boca Raton, FL: CRC Press (1993), chapter 1.
8. M. Pumera, *Electrophoresis*, **29**, 269 (2008).
9. I. Nischang, U. Tallarek, *Electrophoresis*, **28**, 611 (2007).
10. A.D. Zamfira, *J. Chromatogr.*, **1159**, 2 (2007).
11. M. Pumera, *J. Chromatogr. A*, **1113**, 5 (2006).
12. H. Hisamoto, S. Takeda, S. Terabe, *Anal. Bioanal. Chem.*, **386**, 733 (2006).
13. S. Büttgenbach, M. Michalzik, R. Wilke, *Eng. Life Sci.*, **6**, 449 (2006).
14. W.C. Sung, H. Makamba, S.H. Chen., *Electrophoresis*, **26**, 1783 (2005).
15. F. Xu, Y. Baba, *Electrophoresis*, **25**, 2332 (2004).
16. C.J. Evenhuis, R.M. Guijt, M. Macka, P.R. Haddad, *Electrophoresis*, **25**, 3602 (2004).
17. E.A. Doherty, R.J. Meagher, M.N. Albarghouthi, A.E. Barron, *Electrophoresis*, **24**, 34 (2003).
18. K. Huikko, R. Kostiainen, T. Kotiaho, *Eur. J. Pharm. Sci.*, **20**, 149 (2003).
19. A.V. Brocke, G. Nicholson, E. Bayer, *Electrophoresis*, **22**, 1251 (2001).
20. I.M. Lazar, L. Li, Y. Yang, B.L. Karger, *Electrophoresis*, **24**, 3655 (2003).
21. Y. Jin, G.A. Luo, R.J. Wang., *Sepu*, **18**, 313 (2000).
22. F.E. Regnier, B. He, S. Lin, *J. Busse. Trends Biotechnol.*, **17**, 101 (1999).
23. Q. Fang, F.R. Wang, S.L. Wang, S.S. Liu, S.K. Xu, Z.L. Fang, *Anal. Chim. Acta*, **390**, 27 (1999).

24. J. Wang, M.P. Chatrathi., A. Mulchandani, W. Chen, *Anal. Chem.*, **73**, 1804 (2001).

25. J. Wang, M.P. Chatrathi, *Anal. Chem.*, **75**, 525 (2003).

26. D.J. Jackson, J.F. Naber, T.J. Roussel, M.M. Crain, K.M. Walsh, R.S. Keynton, R.P. Baldwin, *Anal. Chem.*, **75**, 3643 (2003).

27. E.M. Smith, H. Xu, A.G. Ewing, *Electrophoresis*, **22**, 363 (2001).

28. R. Kikura-Hanajiri, R.S. Martin, S.M. Lunte, *Anal. Chem.*, **74**, 6370 (2002).

29. E. Guihen, J.D. Glennon, *J. Chromatogr. A*, **1071**, 223 (2005).

30. G. Ping, B. Zhu, M. Jabasini, F. Xu, H. Oka, H. Sugihara, Y. Baba, *Anal. Chem.*, **77**, 7282 (2005).

31. T. Sikanen, S. Tuomikoski, R.A. Ketola, R. Kostiainen, S. Franssila, T. Kotiaho, *Anal. Chem.*, **79**, 9135 (2007).

32. Y. Shi, P.C. Simpson, J.R. Scherer, D. Wexler, C. Skibola, M.T. Smith, R.A. Mathies, *Anal. Chem.*, **71**, 5354 (1999).

33. B.M. Paegel, C.A. Emrich, G.J. Weyemayer, J.R. Scherer, R.A. Mathies, *Proc. Natl. Acad. Sci. USA*, **99**, 574 (2002).

34. A.T. Woolley, R.A. Mathies, *Proc. Natl. Acad. Sci. USA*, **91**, 11348 (1994).

35. A.T. Woolley, G.F. Sensabaugh, R.A. Mathies, *Anal. Chem.*, **69**, 2181 (1997).

36. P.C. Simpson, D. Roach, A.T. Woolley, T. Thorsen, R. Johnston, G.F. Sensabaugh, R.A. Mathies, *Proc. Natl. Acad. Sci. USA*, **95**, 2256 (1998).

37. C.A. Emrich, H.J. Tian, I.L. Medintz, R.A. Mathies, *Anal. Chem.*, **74**, 5076 (2002).

38. C.T. Culbertson, S.C. Jacobson, J.M. Ramsey, *Anal. Chem.*, **70**, 3781 (1998).

39. F. Omasu, Y. Nakano, T. Ichiki, *Electrophoresis*, **26**, 1163 (2005).

40. C.L. Colyer, S.D. Mangru, D.J. Harrison, *J. Chromatogr. A*, **781**, 271 (1997).

41. N. Chiem, D.J. Harrison, *Anal. Chem.*, **69**, 373 (1997).

42. I. Rodriguez, Y. Zhang, H.K. Lee, S.F. Li., *J. Chromatogr. A*, **781**, 287 (1997).

43. V. Linder, E. Verpoorte, N.F. de Rooij, H. Sigrist, W. Thormann, *Electrophoresis*, **23**, 740 (2002).

44. K. Forrer, S. Hammer, B. Helk, *Anal. Biochem.*, **334**, 81 (2004).

45. N.H. Chiem, D.J. Harrison, *Electrophoresis*, **19**, 3040 (1998).

46. S.K. Hsiung, C.H. Lin, G.B. Lee, *Electrophoresis*, **26**, 1122 (2005).

47. S.H. Chen, Y.H. Lin, L.Y. Wang, C.C. Lin, G.B. Lee, *Anal. Chem.*, **74**, 5146 (2002).

48. G. Zhuang, Q. Jin, J. Liu, H. Cong, K. Liu, J. Zhao, M. Yang, H. Wang, *Biomed. Microdevices*, **8**, 255 (2006).

49. T. Sikanen, L. Heikkilä, S. Tuomikoski, R.A. Ketola, R. Kostiainen, S. Franssila, T. Kotiaho, *Anal. Chem.*, **79**, 6255 (2007).

50. Y. Liu, R.S. Foote, S.C. Jacobson, R.S. Ramsey, J.M. Ramsey, *Anal. Chem.*, **72**, 4608 (2000).

51. G.E. Yue, M.G. Roper, C. Balchunas, A. Pulsipher, J.J. Coon, J. Shabanowitz, D.F. Hunt, J.P. Landers, J.P. Ferrance, *Anal. Chim. Acta*, **564**, 116 (2006).

52. J. Li, J.F. Kelly, I. Chernushevich, D.J. Harrison, P. Thibault, *Anal. Chem.*, **72**, 599 (2000).

53. Z. Deyl, I. Miksík, A. Eckhardt, *J. Chromatogr. A*, **990**, 153 (2003).

54. S.W. Tsai, M. Loughran, H. Suzuki, I. Karube, *Electrophoresis*, **25**, 494 (2004).

55. H.K. Musyimi, J. Guy, D.A. Narcisse, S.A. Soper, K.K. Murray, *Electrophoresis*, **26**, 4703 (2005).

56. H. Nagata, M. Tabuchi, K. Hirano, Y. Baba, *Electrophoresis*, **26**, 2687 (2005).

57. J.A. Fruetel, R.F. Renzi, V.A. Vandernoot, J. Stamps, B.A. Horn, J.A. West, S. Ferko, R. Crocker, C.G. Bailey, D. Arnold, B. Wiedenman, W.Y. Choi, D. Yee, I. Shokair, E. Hasselbrink, P. Paul, D. Rakestraw, D. Padgen, *Electrophoresis*, **26**, 1144 (2005).

58. A.P. Dahlin, M. Wetterhall, G. Liljegren, S.K. Bergström, P. Andrén, L. Nyholm, K.E. Markides, J. Bergquist, *Analyst*, **130**, 193 (2005).

59. Y. Sun, M. Lu, X.F. Yin, X.G. Gong, *J. Chromatogr. A*, **1135**, 109 (2006).

60. S.K. Mohanty, D. Kim, D.J. Beebe, *Electrophoresis*, **27**, 3772 (2006).

61. A.W. Chow, *Methods Mol. Biol.*, **339**, 145 (2006).

62. L.H. Silvertand, E. Machtejevas, R. Hendriks, K.K. Unger, W.P. van Bennekom, G.J. de Jong, *J. Chromatogr. B Analyt. Technol. Biomed. Life Sci.*, **839**, 68 (2006).

63. T.M. Phillips, E. Wellner, *J. Chromatogr. A*, **1111**, 106 (2006).

64. B. Zhang, F. Foret, B.L. Karger, *Anal. Chem.*, **72**, 1015 (2000).

65. E. Vasilyeva, J. Woodard, F.R. Taylor, M. Kretschmer, H. Fajardo, Y. Lyubarskaya, K. Kobayashi, A. Dingley, R. Mhatre, *Electrophoresis*, **25**, 3890 (2004).

66. R.F. Renzi, J. Stamps, B.A. Horn, S. Ferko, V.A. Vandernoot, J.A. West, R. Crocker, B. Wiedenman, D. Yee, J.A. Fruetel, *Anal. Chem.*, **77**, 435 (2005).

67. C.W. Kan, C.P. Fredlake, E.A. Doherty, A.E. Barron, *Electrophoresis*, **25**, 3564 (2004).

68. Y.W. Lin, M.F. Huang, H.T. Chang, *Electrophoresis*, **26**, 320 (2005).

69. I. Kheterpal, J.R. Scherer, S.M. Clark, A. Radhakrishnan, J.Y. Ju, C.L. Ginther, G.F. Sensabaugh, R.A. Mathies, *Electrophoresis*, **17**, 1852 (1996).

70. J.R. Scherer, I. Kheterpal, A. Radhakrishnan, W.W. J. Richard, A. Mathies, *Electrophoresis*, **20**, 1508 (1999).

71. J.R. Scherer, B.M. Paegel, G.J. Wedemayer, C.A. Emrich, J. Lo, I.L. Medintz, R.A. Mathies, *Bio. Techniques*, **31**, 1150 (2001).

72. B.M. Paegel, C.A. Emrich, G.J. Weyemayer, J.R. Scherer, R.A. Mathies, *Proc. Natl. Acad. Sci. USA*, **99**, 574 (2002).

73. B.M. Paegel, L.D. Hutt, P.C. Simpson, R.A. Mathies, *Anal. Chem.*, **72**, 3030 (2000).

74. R.J. Meagher, J.A. Coyne, C.N. Hestekin, T.N. Chiesl, R.D. Haynes, J.I. Won, A.E. Barron, *Anal. Chem.*, **79**, 1848 (2007). Epub 2007.

75. J. Chowdhury, G.V. Kagiala, S. Pushpakom, J. Lauzon, A. Makin, A. Atrazhev, A. Stickel, W.G. Newman, C.J. Backhouse, L.M. Pilarski, *J. Mol. Diagn.*, **9**, 521 (2007).

76. S. Liu, H. Ren, Q. Gao, D.J. Roach, R.T. Loder, Jr., T.M. Armstrong, Q. Mao, I. Blaga, D.L. Barker, S.B. Jovanovich, *Proc. Natl. Acad. Sci. USA*, **97**, 5369 (2000).

77. J.W. Hong, K. Hosokawa, T. Fujii, M. Seki, I. Endo, *Biotechnol. Prog.*, **17**, 958 (2001).

78. J.W. Hong, T. Fujii, M. Seki, T. Yamamoto, I. Endo, *Electrophoresis*, **22**, 328 (2001).

79. T. Tang, M.Y. Badal, G. Ocvirk, W.E. Lee, D.E. Bader, F. Bekkaoui, D.J. Harrison, *Anal. Chem.*, **74**, 725 (2002).

80. Z.Q. Xu, T. Hirokawa, T. Nishine, A. Arai, *J. Chromatogr. A*, **990**, 53 (2003).

81. D. Liu, M. Shi, H. Huang, Z. Long, X. Zhou, J. Qin, B. Lin, *J. Chromatogr. B Analyt. Technol. Biomed. Life Sci.*, **844**, 32 (2006).

82. M. Kataoka, S. Inoue, K. Kajimoto, Y. Sinohara, Y. Baba., *Eur. J. Biochem.*, **271**, 2241 (2004).

83. C.H. Lin, G.B. Lee, L.M. Fu, S.H. Chen, *Biosens. Bioelectron.*, **20**, 83 (2004).

84. P.J. Obeid, T.K. Christopoulos, P.C. Ioannou, *Electrophoresis*, **25**, 922 (2004).

85. J. Posedi, M. Drögemüller, T. Schnieder, J. Höglund, J.R. Lichtenfels, G. von Samson-Himmelstjerna, *Parasitol. Res.*, **92**, 421 (2004).

86. J.J. Dooley, H.D. Sage, M.A. Clarke, H.M. Brown, S.D. Garrett, *J. Agric. Food Chem.*, **53**, 3348 (2005).

87. S. Spaniolas, S.T. May, M.J. Bennett, G.A. Tucker, *J. Agric. Food Chem.*, **54**, 7466 (2006).

88. L.M. Fu, C.H. Lin, *Electrophoresis*, **25**, 3652 (2004).

89. Z. Xu, T. Nishine, A. Arai, T. Hirokawa, *Electrophoresis*, **25**, 3875 (2004).

90. Z. Chen, M.A. Burns, *Electrophoresis*, **26**, 4718 (2005).

91. V.J. Sieben, C.J. Backhouse, *Electrophoresis*, **26**, 4729 (2005).

92. P. Hawtin, I. Hardern, R. Wittig, J. Mollenhauer, A. Poustka, R. Salowsky, T. Wulff, C. Rizzo, B. Wilson, *Electrophoresis*, **26**, 3674 (2005).

93. T. Kawabata, M. Watanabe, K. Nakamura, S. Satomura, *Anal. Chem.*, **77**, 5579 (2005).

94. D.K. Kim, S.H. Kang, *J. Chromatogr. A*, **1064**, 121 (2005).

95. T.N. Chiesl, W. Shi, A.E. Barron, *Anal. Chem.*, **77**, 772 (2005).

96. Y.J. Chuang, J.W. Huang, H. Makamba, M.L. Tsai, C.W. Li, S.H. Chen, *Electrophoresis*, **27**, 4158 (2006).

97. R.G. Blazej, P. Kumaresan, R.A. Mathies, *Proc. Natl. Acad. Sci. USA*, **103**, 7240 (2006).

98. F.C. Huang, Y.F. Chen, G.B. Lee, *Electrophoresis*, **28**, 1130 (2007).

99. V. Dolnik, S. Liu., *J. Separation Sci.*, **28**, 1994 (2005).

100. C. Zhang, J. Xu, W. Ma, W. Zheng, *Biotechnol. Adv.*, **24**, 243 (2006).

101. L.A. Legendre, J.M. Bienvenue, M.G. Roper, J.P. Ferrance, J.P. Landers, *Anal. Chem.*, **78**, 1444 (2006).

102. T. Nojima, S. Kaneda, T. Fujii, *Nucleic Acids Symp. Ser. (Oxford)*, **51**, 87 (2007).

103. S. Lutz-Bonengel, T. Sänger, M. Heinrich, U. Schön, U. Schmidt, *Int. J. Legal Med.*, **121**, 68 (2007).

104. J. Qin, Z. Liu, D. Wu, N. Zhu, X. Zhou, Y. Fung, B. Lin, *Electrophoresis*, **26**, 219 (2005).

105. H.J. Tian, A. Jaquins Gerstl, N. Munro, M. Trucco, L. C. Brody, J.P. Landers, *Genomics*, **63**, 25 (2000).

106. J.S. Buch, F. Rosenberger, D. DeVoe, C. Lee, *Micro total analysis systems*, Dordrecht: Kluwer Academic (2002), 233–235.

107. D. Schmalzing, A. Belenky, M.A. Novotny, L. Koutny, O. Salas-Solano, S. El-Difrawy, A. Adourian, *Nucleic Acids Res.*, **28**, 43 (2000).

108. A. Russom, H. Andersson, P. Nilsson, A. Ahmadian, G. Stemme, *Micro total analysis systems*, Dordrecht: Kluwer Academic (2002), 218–220.

109. P. Liu, W.L. Xing, D. Liang, G.-L. Huang, J. Cheng, *Micro total analysis systems*, Dordrecht: Kluwer Academic (2002), 311–313.

110. H.D. Zhang, J. Zhou, Z.R. Xu, J. Song, J. Dai, J. Fang, Z.L. Fang, *Lab. Chip*, **7**, 1162 (2007).

111. M. Pumera, *Electrophoresis*, **28**, 2113 (2007).

112. D.J. Harrison, K. Fluri, K. Seiler, Z. Fan, C.S. Effenhauser, A. Manz, *Science*, **261**, 895 (1993).

113. C.X. Zhang, A. Manz, *Anal. Chem.*, **75**, 5759 (2003).

114. L. Zhang, X.F. Yin, *J. Chromatogr. A*, **1137**, 243 (2006).

115. H. Wu, A. Wheeler, R.N. Zare, *Proc. Natl. Acad. Sci. USA*, **101**, 12809 (2004).

116. Y. Mourzina, A. Steffen, D. Kalyagin, R. Carius, A. Offenhäusser, *Electrophoresis*, **26**, 1849 (2005).

117. E.M. Abad-Villar, P. Kubán, P.C. Hauser, *Electrophoresis*, **26**, 3609 (2005).

118. J.L. Fu, Q. Fang, T. Zhang, X.H. Jin, Z.L. Fang, *Anal. Chem.*, **78**, 3827 (2006).

119. Z. Shen, X. Liu, Z. Long, D. Liu, N. Ye, J. Qin, Z. Dai, B. Lin, *Electrophoresis*, **27**, 1084 (2006).

120. X. Sun, B.A. Peeni, W. Yang, H.A. Becerril, A.T. Woolley, *J. Chromatogr. A*, **1162**, 162 (2007).

121. H. Shadpour, M.L. Hupert, D. Patterson, C. Liu, M. Galloway, W. Stryjewski, J. Goettert, S.A. Soper, *Anal. Chem.*, **79**, 870 (2007).

122. J.G. Shackman, M.S. Munson, D. Ross. *Anal. Chem.*, **79**, 565 (2007).

123. L. Ceriotti, N.F. de Rooij, E. Verpoorte, *Anal. Chem.*, **74**, 639 (2002).

124. X. Liu, X. Liu, A. Liang, Z. Shen, Y. Zhang, Z. Dai, B. Xiong, B. Lin, *Electrophoresis*, **27**, 3125 (2006).

125. C.E. Hop, *Curr. Drug Metab.*, **7**, 557 (2006).

126. A. Ramseier, F. von Heeren, W. Thormann, *Electrophoresis*, **19**, 2967 (1998).

127. N.P. Beard, A.J. de Mello, *Electrophoresis*, **23**, 1722 (2002).

128. N.P. Beard, J.B. Edel, A.J. de Mello, *Electrophoresis*, **25**, 2363 (2004).

129. M.A. Schwarz, *Electrophoresis*, **25**, 1916 (2004).

130. S.J. Varjo, M. Ludwig, D. Belder, M.L. Riekkola, *Electrophoresis*, **25**, 1901 (2004).

131. Y. Deng, H. Zhang, J. Henion, *Anal. Chem.*, **73**, 1432 (2001).

132. Y. Liu, J.A. Vickers, C.S. Henry, *Anal. Chem.*, **76**, 1513 (2004).

133. J.H. Kim, C.J. Kang, Y.S. Kim, *Biosens. Bioelectron.*, **20**, 2314 (2005).

134. M. Johirul, A. Shiddiky, R.E. Kim, Y.B. Shim, *Electrophoresis*, **26**, 3043 (2005).

135. P. Schulze, M. Ludwig, F. Kohler, D. Belder, *Anal. Chem.*, **77**, 1325 (2005).

136. A. Schuchert-Shi, P. Kubáň, P.C. Hauser, *Electrophoresis*, **28**, 4690 (2007).

137. H. Hu, Q. Xiong, C. Li, H. Gao, Q. Yang, Z. Liang, *Zhonghua Yi Xue Yi Chuan Xue Za Zhi*, **24**, 709 (2007).

138. Y. Murakami, T. Morita, T. Kanekiyo, E. Tamiya, *Biosens. Bioelectron.*, **16**, 1009 (2001).

139. J. Wang, M.P. Chatrathi, A. Ibañez, *Analyst*, **126**, 1203 (2001).

140. J. Wang, M.P. Chatrathi, B. Tian, *Anal. Chem.*, **73**, 1296 (2001).

141. D.E. Starkey, Y. Abdelaziez, C.H. Ahn, J. Tu, L. Anderson, K.R. Wehmeyer, N.J. Izzo, A.N. Carr, K.G. Peters, J.J. Bao, H.B. Halsall, W.R. Heineman, *Anal. Biochem.*, **316**, 181 (2003).

142. G.S. Zhuang, J. Liu, C.P. Jia, Q.H. Jin, J.L. Zhao, H.M. Wang, *J. Separation Sci.*, **30**, 1350 (2007).

143. J.F. Dishinger, R.T. Kennedy, *Anal. Chem.*, **79**, 947 (2007).

144. M.G. Roper, J.G. Shackman, G.M. Dahlgren, R.T. Kennedy, *Anal. Chem.*, **75**, 4711 (2003).

145. Y. Xu, M.W. Little, K.K. Murray, *J. Am. Soc. Mass. Spectrom.*, **17**, 469 (2006).

146. T.G. Henares, M. Takaishi, N. Yoshida, S. Terabe, F. Mizutani, R. Sekizawa, H. Hisamoto, *Anal. Chem.*, **79**, 908 (2007).

147. L. Licklider, X.Q. Wang, A. Desai, Y.C. Tai, T.D. Lee, *Anal. Chem.*, **72**, 367 (2000).

148. Y.T. Wu, Y.C. Chen, *Anal. Chem.*, **77**, 2071 (2005).

149. Y.Y. Ling, X.F. Yin, Z.L. Fang, *Electrophoresis*, **26**, 4759 (2005).

150. J. Qin, N. Ye, L. Yu, D. Liu, Y. Fung, W. Wang, X. Ma, B. Lin, *Electrophoresis*, **26**, 1155 (2005).

151. E.X. Vrouwe, R. Luttge, I. Vermes, A. van den Berg, *Clin. Chem.*, **53**, 117 (2007).

152. Z. Zhuang, J.A. Starkey, Y. Mechref, M.V. Novotny, S.C. Jacobson, *Anal. Chem.*, **79**, 7170 (2007).

153. E. Maeda, M. Kataoka, M. Hino, K. Kajimoto, N. Kaji, M. Tokeshi, J. Kido, Y. Shinohara, Y. Baba, *Electrophoresis*, **28**, 2927 (2007).

154. T.M. Phillips, *Electrophoresis*, **25**, 1652 (2004).

155. J. Gao, X.F. Yin, Z.L. Fang, *Lab. Chip*, **4**, 47 (2004).

156. E.X. Vrouwe, P. Koelling, R. Luttge, A. van den Berg, *Nederlands Tijdschrift Voor Klinische Chemie En aboratoriumgeneeskunde*, **29**, 295 (2004).

157. Y. Du, J. Yan, W. Zhou, X. Yang, E. Wang, *Electrophoresis*, **25**, 3853 (2004).

158. Y. Deng, J. Henion, J. Li, P. Thibault, C. Wang, D.J. Harrison, *Anal. Chem.*, **73**, 639 (2001).

159. J. Wang, M.P. Chatrathi, B. Tian, R. Polsky, *Anal. Chem.*, **72**, 2514 (2000).

160. O.T. Chan, D.A. Herold, *Clin. Chem.*, **52**, 2141 (2006).

161. R. Wilke, S. Buttgenbach, *Biosens. Bioelectron*, **19**, 149 (2003).

162. M. Žuborova, M. Masr, D. Kaniansky, M. Johnck, B. Stanislawski, *Electrophoresis*, **23**, 774 (2002).

163. J. C. Fanguy, C. S. Henry, *Electrophoresis*, **23**, 767 (2002).

164. S. Suzuki, N. Shimotsu, S. Honda, A. Arai, H. Nakanishi, *Electrophoresis*, **22**, 4023 (2001).

165. K. Usui, A. Hiratsuka, K. Shiseki, Y. Maruo, T. Matsushima, K. Takahashi, Y. Unuma, K. Sakairi, I. Namatame, Y. Ogawa, K. Yokoyama, *Electrophoresis*, **27**, 3635 (2006).

166. Y. Zhang, G. Ping, B. Zhu, N. Kaji, M. Tokeshi, Y. Baba, *Electrophoresis*, **28**, 414 (2007).

167. M. Masár, M. Danková, E. Olvecká, A. Stachurová, D. Kaniansky, B. Stanislawski, *J. Chromatogr. A*, **1084**, 101 (2005).

168. M. Masár, K. Poliakova, M. Dankova, D. Kaniansky, B. Stanislawski, *J. Separation Sci.*, **28**, 905 (2005).

169. A.M. Skelley, R.A. Mathies, *J. Chromatogr. A*, **1132**, 304 (2006).

170. C.N. Jayarajah, A.M. Skelley, A.D. Fortner, R.A. Mathies, *Anal. Chem.*, **79**, 8162 (2007).

171. J. Borowsky, G.E. Collins, *Analyst*, **132**, 958 (2007).

172. D.W. Armstrong, J.M. Scneiderheinze, J.P. Kullman, L. He, *Microbiol. Lett.*, **194**, 33 (2001).

173. R.C. Ebersole, R.M. McCormick, *Bio/Technology*, **11**, 1278 (1993).

174. M.A. Rodriguez, D.W. Armstrong, *J. Chromatogr. B*, **800**, 7 (2004).

175. X. Zhou, D. Liu, R. Zhong, Z. Dai, D. Wu, H. Wang, Y. Du, Z. Xia, L. Zhang, X. Mei, B. Lin, *Electrophoresis*, **25**, 3032 (2004).

176. F.C. Huang, C.S. Liao, G.B. Lee, *Electrophoresis*, **27**, 3297 (2006).

177. N.C. Tsai, C.Y. Sue, *Biosens. Bioelectron.*, **22**, 313 (2006).

178. V. Kolivoška, V.U. Weiss, L. Kremser, B. Gas, D. Blaas, E. Kenndler, *Electrophoresis*, **28**, 4734 (2007).

179. V.U. Weiss, V. Kolivoska, L. Kremser, B. Gas, D. Blaas, E. Kenndler, *J. Chromatogr. B Anal. Technol. Biomed. Life Sci.*, **860**, 173 (2007).

180. E.T. Lagally, J.R. Scherer, R.G. Blazej, N.M. Toriello, B.A. Diep, M. Ramchandani, G.F. Sensabaugh, L.W. Riley, R.A. Mathies, *Anal. Chem.*, **76**, 3162 (2004).

181. A.W. Lantz, Y. Bao, D.W. Armstrong, *Anal. Chem.*, **79**, 1720 (2007).

182. M. Pumera, *Electrophoresis*, **27**, 244 (2006).

183. J. Wang, M. Pumera, G. Collins, F. Opekar, I. Jelínek, *Analyst*, **127**, 719 (2002).

184. J. Wang, A. Escarpa, M. Pumera, J. Feldman, *J. Chromatogr. A*, **952**, 249 (2002).

185. J. Wang, M. Pumera, *Anal. Chem.*, **74**, 5919 (2002).

186. A. Bromberg, R.A. Mathies, *Anal. Chem.*, **75**, 1188 (2003).

187. A. Bromberg, R.A. Mathies, *Electrophoresis*, **25**, 1895 (2004).

188. I. Ali, H.Y. Aboul-Enein, *Chiral pollutants: Distribution, toxicity and analysis by chromatography and capillary electrophoresis*, Chichester: Wiley (2004).

189. I. Ali, H.Y. Aboul-Enein, *Instrumental methods in metal ions speciation: Chromatography, capillary electrophoresis and electrochemistry*, New York: Taylor & Francis (2006).

190. J. Wang, M. Pumera, G.E. Collins, A. Mulchandani, *Anal. Chem.*, **74**, 6121 (2002).

191. Q. Lu, G.E. Collins, T. Evans, M. Hammond, J. Wang, A. Mulchandani, *Electrophoresis*, **25**, 116 (2004).

192. F. Li, D.D. Wang, X.P. Yan, R.G. Su, J.M. Lin, *J. Chromatogr. A*, **1081**, 232 (2005).

193. F. Li, D.D. Wang, X.P. Yan, J.M. Lin, R.G. Su, *Electrophoresis*, **26**, 226 (2005).

194. O. Vogt, M. Pfister, U. Marggraf, A. Neyer, R. Hergenroder, P. Jacob, *Lab. Chip*, **5**, 205 (2005).

195. A.Y.N. Hui, G. Wang, B. Linb, W.T. Chan, *J. Anal. At. Spectrom.*, **21**, 134 (2006).

196. J.P. Kutter, R.S. Ramsey, S.C. Jacobson, J.M. Ramsey, *J. Microcol. Sep.*, **10**, 313 (1998).

197. V. Rohlícek, Z. Deyl, *J. Chromatogr. B Anal. Technol. Biomed. Life Sci.*, **770**, 19 (2002).

198. J. Wang, W. Siangproh, S. Thongngamdee, O. Chailapakul, *Analyst*, **130**, 1390 (2005).

199. T. Le Saux, H. Hisamoto, S. Terabe, *J. Chromatogr. A*, **1104**, 352 (2006).

200. M. Fleger, D. Siepe, A. Neyer, *IEE Proc. Nanobiotechnol.*, **151**, 159 (2004).

201. J.B. Edel, N.P. Beard, O. Hofmann, J.C. deMello, D.D. Bradley, A.J. deMello, *Lab. Chip*, **4**, 136 (2004).

202. H.T. Feng, H.P. Wei, S.F. Li., *Electrophoresis*, **25**, 909 (2004).

203. I. Nikcevic, S.H. Lee, A. Piruska, C.H. Ahn, T.H. Ridgway, P.A. Limbach, K.R. Wehmeyer, W.R. Heineman, C.J. Seliskar, *J. Chromatogr. A*, **1154**, 444 (2007).

204. T. Ujiie, T. Kikuchi, T. Ichiki, Y. Horiike, *Jpn. J. Appl. Phys.*, **39**, 3677 (2000).

205. S. Wakida, A. Chiba, T. Matsuda, K. Fukushi, H. Nakanishi, X. Wu, H. Nagai, S. Kurosawa, S. Takeda, *Electrophoresis*, **22**, 3505 (2001).

206. K.S. Shin, Y.H. Kim, J. Ah-Min, S.M. Kwak, S.K. Kima, E.G. Yang, J.H. Park, B.K. Ju, T.S. Kim, J.Y. Kang, *Anal. Chim. Acta.*, **573**, 164 (2006).

207. M. Brivio, R.E. Oosterbroek, W. Verboom, A. van den Berg, D.N. Reinhoudt, *Lab. Chip*, **5**, 1111 (2005).

208. A.R. Stettler, M.A. Schwarz, *J. Chromatogr. A*, **1063**, 217 (2005).

209. A.B. Fuchs, A. Romani, D. Freida, G. Medoro, M. Abonnenc, L. Altomare, I. Chartier, D. Guergour, C. Villiers, P.N. Marche, M. Tartagni, R. Guerrieri, F. Chatelain, N. Manaresi, *Lab. Chip*, **6**, 121 (2006).

210. T.G. Henares, S. Funano, S. Terabe, F. Mizutani, R. Sekizawa, H. Hisamoto, *Anal. Chim. Acta.*, **589**, 173 (2007).

211. J. Cheng, M.A. Shoffner, K.R. Mitchelson, L.J. Kricka, P. Wilding, *J. Chromatogr. A*, **732**, 151 (1996).

212. K.T. Liao, C.M. Chenb, H.J. Huanga, C.H Lin, *J. Chromatogr. A*, **1165**, 213 (2007).

213. B.F. Liu, *Anal. Chem.*, **75**, 36 (2003).

214. X.J. Huang, Q.S. Pu, Z.L. Fang, *Analyst*, **126**, 281 (2001).

215. J. Tanyanyiwa, S. Leuthardt, P.C. Hauser, *Electrophoresis*, **23**, 3659 (2002).

216. G. Shekhawat, S.H. Tark, V.P. Dravid, *Science*, **311**, 1592 (2006).

217. G. Wu, R.H. Datar, K.M. Hansen, T. Thundat, R.J. Cote, A. Majumdar, *Nat. Biotechnol.*, **19**, 856 (2001).

218. C.A. Savran, S.M. Knudsen, A.D. Ellington, S.R. Manalis, *Anal. Chem.*, **76**, 3194 (2004).

219. K.W. Wee, H.Y. Kang, J. Park, J.Y. Kang, D.S. Yoon, J.H. Park, T.S. Kim, *Biosens. Bioelectron.*, **20**, 1932 (2005).

220. N. Backmann, C. Zahnd, F. Huber, A. Bietsch, A. Pluckthun, H.P. Lang, H.J. Guntherodt, M. Hegner, C. Gerber, *Proc. Natl. Acad. Sci. USA*, **102**, 14587 (2005).

221. J. Fritz, M.K. Baller, H.P. Lang, H. Rothuizen, P. Vettiger, E. Meyer, H.-J. Guntherodt, C. Gerber, J.K. Gimzewski, *Science*, **288**, 316 (2000).

222. F. Huber, M. Hegner, C. Gerber, H.J. Guntherodt, H.P. Lang, *Biosens. Bioelectron.*, **21**, 1599 (2006).

223. R. Mukhopadhyay, M. Lorentzen, J. Kjems, F. Besenbacher, *Langmuir*, **21**, 8400 (2005).

224. R. McKendry, J. Zhang, Y. Arntz, T. Strunz, M. Hegner, H.P. Lang, M.K. Baller, U. Certa, E. Meyer, H.J. Guntherodt, C. Gerber, *Proc. Natl. Acad. Sci. USA*, **99**, 9783 (2002).

225. H. Shen, Q. Fang, Z.L. Fang. *Lab. Chip*, **6**, 1387 (2006).

CHAPTER 9

CHIRAL SEPARATIONS BY NANOLIQUID CHROMATOGRAPHY AND NANOCAPILLARY ELECTROPHORESIS

9.1. INTRODUCTION

The history of enantiomeric resolution starts with the discovery of chirality in 1809 by Haüy [1] and Pasteur's experiments on different destruction rates of *dextro* and *levo* ammonium tartarate by the mold *Penicillium glaucum* [2]. Now, it has been established that one of the enantiomers may be toxic or inactive or ballast. About 68% of clinically prescribed drugs are being sold as racemates [3,4]. To ensure the desired optimum therapeutic effect, it appears convenient to administer the eutomer. However, applying a single enantiomer to humans does not necessarily prevent side effects or tissue/organ damage as, sometimes, chiral inversion or racemization, can occur *in vitro*. An example of a chiral inversion without negative side effects is ibuprofen, where the inactive (R)-$(-)$-enantiomer is converted into the active (S)-$(+)$-form by an enzymatic mechanism [5]. On the other hand, $(+)$-thalidomide racemizes into the harmful $(-)$-antipode *in vivo*, leading to malformations of embryos in pregnant women [3,6].

Due to the importance of chiral drugs and pharmaceuticals, Witte et al. [7] and Rauws and Groen [8] reviewed the status of regulatory aspects of chiral medicinal products in the pharmaceutical industries of the United States, Japan, and some European countries. The U.S. Food and Drug Administration (FDA) and other

Nanochromatography and Nanocapillary Electrophoresis. By Ali, Aboul-Enein, and Gupta
Copyright © 2009 John Wiley & Sons, Inc.

authorities have put forward certain regulations on the marketing of racemic drugs and pharmaceuticals, which have resulted in an increasing demand for chiral separation methods.

In environmental science chirality has an important role as degradation of some achiral pollutants result in chiral toxic metabolites. Therefore, predicting the exact toxicities of the pollutant concentrations of both enantiomers is required and essential. For example, two enantiomers of α-hexachlorocyclo-hexane pesticide have different toxicities. Moreover, the rates of degradation of the enantiomers of α-hexachlorocyclohexane are also different [9,10].

Different analytical and preparative methods have been developed [11–15]. Among these, chromatography and capillary electrophoresis are the best choices for this purpose, owing to their wide range of applications, ease of operation, good efficiencies, sensitivities, selectivities, and reproducibilities [11–14]. Due to the development of science and technology during the last few years; the demand for analytical separations at nano or low levels has been met especially in proteomics, genomics, and drug development. This chapter describes chiral separations of some compounds by nanoliquid chromatographic and nanocapillary electrophoretic techniques.

9.2. NANOLIQUID CHROMATOGRAPHY

Certainly, a chiral environment is essential for enantiomeric resolution in chromatography. Therefore, different chiral compounds are required. The most commonly used chiral molecules are polysaccharides, cyclodextrins, macrocyclic glycopeptide antibiotics, proteins, crown ethers, ligand exchangers, Pirkle's-type molecules, and several others [11,12]. These compounds are called chiral selectors and are used either in mobile phases or in stationary phases. The former class is called chiral mobile phase additives (CMPA) while the latter are termed chiral stationary phases (CSPs). Presently, in conventional HPLC, CSPs are well developed and about 98% of enantiomeric resolution is achieved successfully. On the other hand, chiral compounds are used as mobile phase additives in conventional CEC and MEKC. As discussed in earlier chapters, microfluidic devices are not well developed and, hence, no information was found in the literature on chiral separations using CSPs. However, enantiomeric resolution in liquid chromatography has been achieved using the mobile phase additive approach. Pumera [16] reviewed chiral separations using liquid chromatography on microchips. Not much work has been carried out on chiral resolution on microchips. The available literature is described here.

Pumera and coworkers [17] used nano-HPLC for enantiomeric resolution of ephedrine, pseudoephedrine, and ibuprofen by using β-cyclodextrin as chiral selector. Wallenborg et al. [18] reported on-chip chiral separation of amphet-amine and related compounds labeled with 4-fluoro-7-nitrobenzofurazane. The chiral separation of norephedrine, ephedrine, cathinone, pseudoephedrine, methcathinone, amphetamine, and methamphetamine was demonstrated using nanomicellar electrokinetic capillary chromatography (NMEKC) and laser-induced fluorescence (LIF) detection. The neutral and negatively charged cyclodextrins were used as mobile phase additives with and without the addition of an organic modifier. Sodium dodecyl sulfate (SDS) was used as surfactant. The best results were obtained using a highly sulfated γ-cyclodextrin (HS-γ-CD) in combination with a low concentration of SDS. The racemic mixtures of norephedrine, ephedrine, cathinone, amphetamine, pseudoephedrine, meth-cathinone, and methamphetamine were resolved using 50 mM phosphate buffer (pH 7.35), containing 10 mM HS-γ-CD and 1.5 mM SDS as the mobile phase. It is very interesting to observe that the authors attempted to resolve all seven racemates within a single run, which is very difficult and exceptional in chiral resolution. Figure 9.1 demonstrates the enantiomeric sep-aration of these molecules, which clearly indicates complete resolution of all racemates except norephedrine and cathinone. Furthermore, the authors attempted to optimize the chiral separations of these drugs by varying the con-centrations of HS-γ-CD as shown in Fig. 9.2, indicating the best resolution at 20 mM HS-γ-CD. This may be due to the formation of stable inclusion

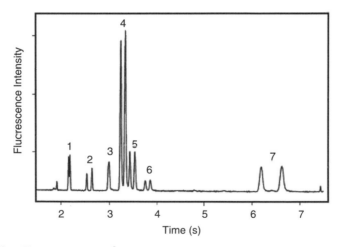

Figure 9.1   Chromatogram of chiral separation of (1) norephedrine, (2) ephedrine, (3) cathinone, (4) amphetamine, (5) pseudoephedrine, (6) methcathinone, and (7) methamphetamine on S-folded separation channel (160 mm length) using 50 mM phosphate buffer (pH 7.35) with 10 mM HS-γ-cyclodextrin and 5 mM SDS at 8 kV/cm potential [18].

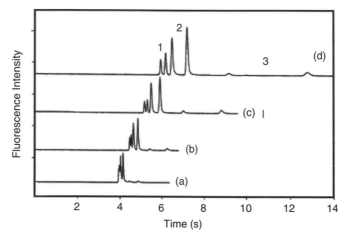

Figure 9.2   Effect of HS-γ-cyclodextrin concentration on the chiral separation of (1) norephedrine, (2) amphetamine, and (3) methamphetamine on S-folded separation channel (160 mm length) using 50 mM phosphate buffer (pH 7.35) with 5 mM SDS at 8 kV/cm potential. (a) 5 mM, (b) 10 mM, (c) 15 mM, and (d) 20 mM HS-γ-cyclodextrins [18].

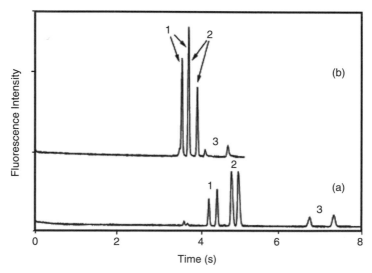

Figure 9.3   Effect of SDS added to the run buffer on the separation of (1) ephedrine, (2) pseudoephedrine, and (3) methamphetamine on S-folded separation channel (145 mm length) using 50 mM phosphate buffer (pH 7.35) with 10 mM HS-γ-cyclodextrin at 8 kV/cm potential. (a) 10 mM HS-γ-CD and (b) 10 mM HS-γ-CD [18].

complexes at this concentration. Besides, the retention time was quite high due to the charged nature of CD. Therefore, the value of the zeta potential decreased with consequent effect. Similarly, Fig. 9.3 represents the presence and absence of SDS on the chiral resolution of these species; the best concentration was achieved when 1 mM SDS was used. This is due to the formation of micelles, which provide greater chances of inclusion complex formation and various interactions resulting in better chiral resolution.

Rodriguez et al. [19] described NMEKC for the chiral resolution of fluorescein isothiocyanate (FITC)-amino acids. The mobile phase used was 100 mM borate buffer having 10 mM γ-cyclodextrin with 30 mM SDS. The separation was achieved on a 60 μm $\times$ 15 μm $\times$ 7 cm (effective length 4.8 cm) channel. The applied voltages were 1.5 and 30 kV for electrokinetic injection and separation, respectively. Similarly, Zeng et al. [20] described chiral separations of FITC-labeled amino acids with gel NCEC on a polydimethylsiloxane (PDMS) microfluidic device. Six pairs of FITC-labeled dansyl amino acids (Dns-AAs) were separated in a 36 mm effectual separation channel in less than 120 seconds. Allyl-γ-cyclodextrin was used as a chiral selector and cross-linker was bonded in γ-CD bonded polyacrylamide gel, which was the separation media, and was immobilized in a PDMS microchannel through the stable linkage of 3-(trimethoxysilyl)-propyl methacrylate. Furthermore, the authors described the longevity of the PDMS microfluidic device.

9.3. NANOCAPILLARY ELECTROPHORESIS

Contrary to conventional HPLC, almost 98% of chiral resolution in CE is carried out using the chiral selector as a mobile phase additive. Again all the common chiral selectors used in NLC can also be used in NCE. But, unfortunately, few chiral molecules have been tested in NCE for enantiomeric resolution of some racemates. To the best of our knowledge only cyclodextrins and protein-based chiral mobile phase additives have been used for this purpose. Manz and coworkers discussed chiral separations by NCE in their reviews in 2004 [21] and 2006 [22]. Later on, Pumera [16] reviewed the use of microfluidic devices for enantiomeric resolutions in capillary electrophoresis. Not much work has been carried out on chiral resolution in NCE but the papers that are available are discussed here.

Pumera and coworkers [17] exploited β-cylodextrin as chiral selector for the enantiomeric resolution of ephedrine, pseudoephedrine, and ibuprofen in NCEC. Male and Luong [23] described chiral analysis of three neurotransmitters (norepinephrine, epinephrine, and isoproterenol) by NCE using 25 mM

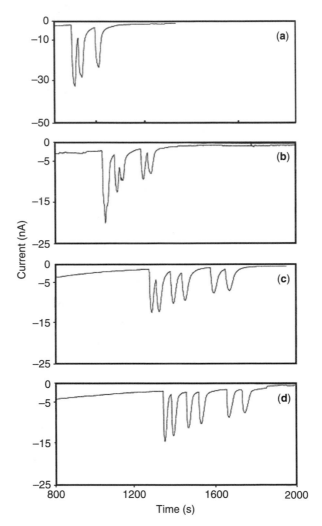

Figure 9.4   Effect of heptakis(2,6,di-O-methyl)-β-cyclodextrin concentration in 50 mM Tris-phosphate (pH 2.5) buffer in NCE separation of neurotransmitters. (a) 0 mM, (b) 5 mM, (c) 15 mM, and (d) 25 mM [29].

heptakis(2,6,di-O-methyl)-β-cyclodextrin as chiral selector in 50 mM Tris-phosphate (pH 2.5) buffer. Furthermore, the resolution was optimized by using various concentrations of cyclodextrin as shown in Fig. 9.4 and it was observed that the order of separation was a > b > c > d with best resolution at 25 mM concentration of cyclodextrin. Similarly, Schwarz and Hauser [24] described the enantiomeric resolution of adrenaline, noradrenaline, ephedrine, and pseudoephedrine by NCE. The separation of the enantiomers was achieved by employing carboxymethyl-β-cyclodextrin as chiral selector in the buffer, partly with the addition of the crown ether 18-crown-6. The detection was achieved by an amperometric method.

Ludwig et al. [25] reported chiral separations of basic and acidic compounds by using NCE with a UV detector. The racemates studied were 1-phenylethylamine, α-,4-dimethylbenzylamine, N-methyl-1-phenylethylamine, 2-diphenylethylamine, 2-(3-chlorophenoxy)-propionic acid, tropic acid, N,α-dimethylbenzylamine, pseudoephedrine, and norephedrine. The sulfated cyclodextrins (HS-α-CD, HS-β-CD, and HS-γ-CD) were used as chiral selectors and separation was achieved within 60 seconds. Furthermore, the authors reported enantiomeric excess (ee) values for these compounds. Besides, the authors reported successful separation of racemic mixture of alprenolol, terbutaline, and tocainide and its derivatives such as N-(2,4-dimethylphenyl)alaninamide, 3-amino-N-(2,6-dimethylphenyl)-2-methylpropanamide, N-phenylalaninamide, N-(2-methylphenyl)-alaninamide, N-(2,4,6-trimethylphenyl)-alaninamide and 3-amino-N-(2,6-dimethylphenyl)-butanamide under the reported experimental conditions. Figure 9.5 represents the chiral resolution of three drugs within 11 seconds. The authors claimed the repeatability of this method as shown in Fig. 9.6, which clearly indicates a quite good reproducibility. Some applications of this paper are summarized in Table 9.1, indicating chiral selectors used, retention times, and resolution factors. These workers advocated NCE as a great potential instrument for fast chiral analysis and high throughput screening.

Balss et al. [26] reported enantiomeric resolution of glutamic acid, dansyl-DL-glutamic acid, and baclofen with chiral temperature gradient focusing in

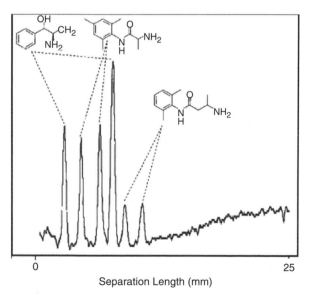

Separation Length (mm)

Figure 9.5 Electropherogram of chiral separation of three basic drugs [25].

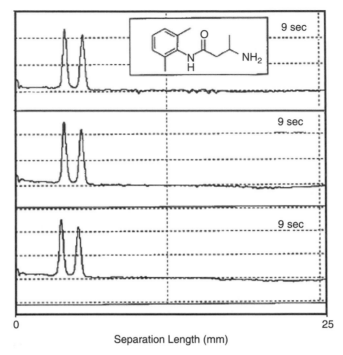

Figure 9.6   Reproducibility of chiral separation of tocainide derivative in NCE [25].

NCE. Enantiomeric separations were accomplished by the addition of α- and β-cyclodextrins (chiral selectors), which caused the different enantiomers of an analyte to focus at different positions along a microchannel. According to the authors, the method showed high performance and facile reversal of peak order, desirable for analysis of trace enantiomeric impurities. Furthermore, they claimed this method provided 1000-fold concentration enhancement in comparison to conventional CE, with improved limits of detection. The optimized chiral separation of dansyl-DL-glutamic acid is shown in Fig. 9.7 under varied experimental conditions. This figure indicates a good resolution due to reducing steepness of the temperature gradient.

Gao et al. [27] described enantiomeric resolution of FITC-labeled basic drugs (baclofen, norfenefrine, and tocainide) by NCE using several neutral cyclodextrins as chiral selectors. FITC-baclofen enantiomers were separated completely by γ-CD while resolution of FITC-norfenefrine enantiomers was achieved by dimethyl-β-cyclodextrins. Furthermore, the authors studied the feasibility of using one chiral selector to separate multiple racemic samples on a four-channel chip. Figure 9.8 indicates the role of some chiral selectors for enantiomeric resolution of FITC-baclofen and FITC-norfenefrine racemates. A perusal of this figure indicates that HP-α-CD and HP-β-CD are

TABLE 9.1   NCE Data for Chiral Resolution of 19 Drugs[a]

Compound	Chiral Selector	t_{r1}	t_{r2}	R_S
	HS-γ-CD	39	83	2.1
	HS-α-CD	15	28	2.6
	HS-β-CD	13	21	2.1
	HS-γ-CD	4	18	6.4
	HS-α-CD	20	23	1.9
	HS-γ-CD	28	56	2.6
	HS-γ-CD	4	25	7.3
	HS-α-CD	31	70	2.5
	HS-γ-CD	23	37	2.3

(*Continued*)

TABLE 9.1   NCE Data for Chiral Resolution of 19 Drugs[a] (*Continued*)

Compound	Chiral Selector	t_{r1}	t_{r2}	R_S
(structure: N-phenyl 2-aminopropanamide)	HS-γ-CD	11	34	3.8
(structure: N-(2-methylphenyl) 2-aminopropanamide)	HS-α-CD	7	36	6.3
(structure: N-(trimethylphenyl) 2-aminopropanamide)	HS-γ-CD	11	26	4.1
(structure: N-(2,6-dimethylphenyl) 3-aminobutanamide)	HS-γ-CD	9	40	5.7
(structure: 2-(3-chlorophenoxy)propanoic acid)	HS-α-CD	27	43	2.2
(structure: 3-hydroxy-2-phenylpropanoic acid)	HS-β-CD	22	50	2.9
(structure: N-(1-phenylethyl)amine)	HS-γ-CD	58	74	1.7
(structure: 2-phenylpropanoic acid)	HS-γ-CD	31	59	2.3
(structure: ephedrine, OH/CH3/NHCH3)	HS-β-CD	10	16	3.0
(structure: norephedrine, OH/CH3/NH2)	HS-γ-CD	2.5	37	12.3

[a]Ref. [25].

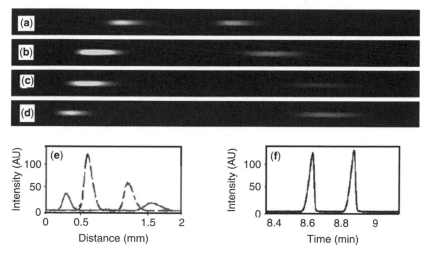

Figure 9.7 Optimization of chiral TGF separation of dansyl-DL-glutamic acid. (a) $T_1 = 13$ to $T_2 = 40°C$, (b) $T_1 = 11$ to $T_2 = 30°C$, (c) $T_1 = 15$ to $T_2 = 30°C$, (d) $T_1 = 2.5$ to $T_2 = 15°C$, (e) intensity vs. distance plots for chiral TGF separations of dansyl-DL-glutamic acid (dotted curve, resolution 2.4, is from (a), complete line curve, resolution 3.8, is from the optimized separation (d), and (f) chiral NCE electropherogram of the same [26].

the best chiral selectors for enantiomeric resolution of FITC-baclofen and FITC-norfenefrine, respectively. Furthermore, the authors reported enantiomeric resolution of FITC-baclofen and FITC-norfenefrine racemates on a two-channel chip as shown in Fig. 9.9, indicating good resolution of both analytes. Besides, the authors presented an excellent comparison of chiral separations of FITC-baclofen, FITC-norfenefrine, and FITC-tocainide along with NCE behavior of FITC as shown in Fig. 9.10 indicating the best resolution of baclofen.

Piehl et al. [28] reported chiral separation of DNS-amino acids in NCE with fluorescence detection. The sulfated cyclodextrins were used as chiral selectors. The separation was achieved within 720 ms by applying 2.01 kV/cm as applied potential. The chiral chromatograms are shown in Fig. 9.11, indicating a good resolution. The retention time and resolution factors of some amino acids are given in Table 9.2. Hutt et al. [29] described enantiomeric resolution of FITC-amino acids within 5 minutes using NCE. The amino acids studied were alanine, glutamine, and asparagine, with carbonate buffer (pH 10.0), containing γ-cyclodextrin, at 550 V/cm applied voltage. Furthermore, the authors optimized chiral separation by adjusting the concentrations of cyclodextrins and temperature. The authors reported 5 mM concentration of

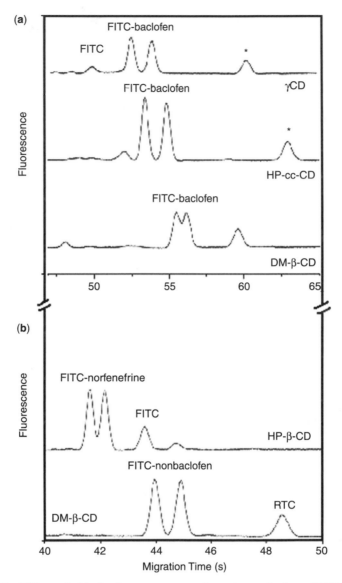

Figure 9.8 Effect of chiral selectors on enantiomeric resolution of FITC-baclofen and FITC-norfenefrine in NCE. (a) CD, HP-α-CD, and DM-β-CD, (b) HP-β-CD and DM-β-CD [27].

γ-cyclodextrin was found to be the best one (Fig. 9.12). Zeng et al. [30] described enantiomeric resolution of FTIC-labeled DNS-amino acids by using allyl-β-cyclodextrin. Nakajima et al. [31] reported chiral separation of 11 NBD-amino acids by ligand exchange-based NCE. A Cu(II) complex with L-prolinamide was used as a chiral selector. The enantiomers of five

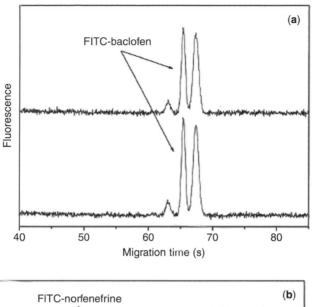

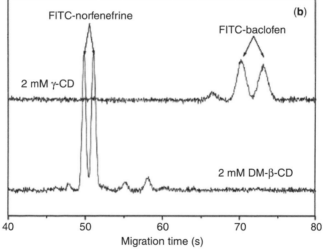

Figure 9.9   Chiral separations of FITC-baclofen and FITC-norfenefrine on a two-channel chip. (a) Parallel analyses of FITC-baclofen in 2 mM γ-CD and (b) chiral separations of FITC-baclofen and FITC-norfenefrine in 2 mM γ-CD and 2 mM DM-γ-CD, respectively [27].

NBD-amino acids (serine, threonine, valanie, phenylalanine, and histidine) were separated by using a 20 mM ammonium acetate buffer (pH 9.0) containing 10 mM copper acetate, 20 mM L-prolinamide, and 1 mM SDS. The elution order was the L-enantiomer followed by the D-form except in NBD-histidine. Weng et al. [32] described chiral separation of tryptophan on a poly(methyl methacrylate) microfluidic chip coated with bovine serum albumin (32 mm

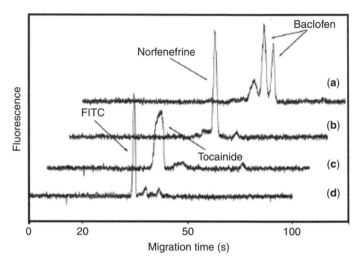

Figure 9.10 A comparison of electropherograms of enantionmeric resolution of (a) FITC-baclofen, (b) FITC-norfenefrine, and (c) FITC-tocainide, and electrophoresis of (d) FITC on the four-channel chip using 20 mM phosphate buffer containing 2 mM γ-CD, pH 9.3.

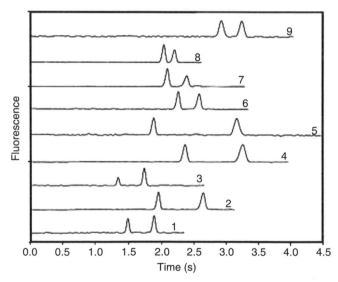

Figure 9.11 Chiral separation of DNS-amino acids using 7 mm separation length and 2.0 kV/cm potential with 25 mM triethylammonium phosphate buffer (pH 2.5) as BGE containing 2% HS-γ-CD as chiral selector. Peaks are (1) DNS-tryptophan, (2) DNS-norleucine, (3) DNS-phenylalanine, (4) DNS-methionine, (5) DNS-aspartic acid, (6) DNS-aminobutyric acid, (7) DNS-leucine, (8) DNS-norvaline, and (9) DNS-glutamic acid [28].

TABLE 9.2   NCE Data for Chiral Separation of DNS-Amino Acids[a]

Compounds	t_r	R_S
DNS-tryptophan	0.73	1.80
DNS-norleucine	0.72	1.60
DNS-phenylalanine	0.82	1.90
DNS-methionine	0.84	1.40
DNS-aspartic acid	0.83	1.80
DNS-aminobutyric acid	1.43	1.60
DNS-leucine	1.40	1.50
DNS-nor-valine	2.24	1.50
DNS-glutamic acid	2.71	1.50

[a]Ref. 25.

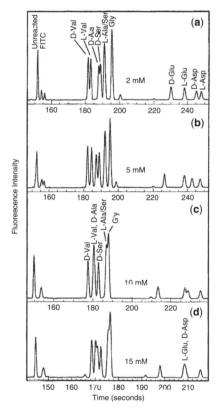

Figure 9.12   Effect of γ-cyclodextrin concentrations on the enantioselectivity of five FITC-labeled amino acids and Gly: (a) 2.0 mM, (b) 5.0 mM, (c) 10.0 mM, and (d) 15.0 mM [29].

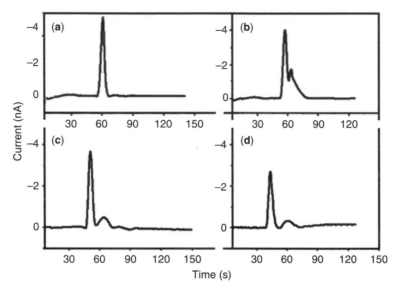

Figure 9.13   Effect of buffer pH on enantiomeric resolution of DL-tryptophan. (a) 5.0, (b) 6.0, (c) 7.0, and (d) 8.0, with 20 mM phosphate buffer as BGE at 0.6 V/cm potential [32].

channel) coupled with laser-induced fluorescence. The separation was achieved within 70 seconds. The background electrolyte (BGE) used was 20 mM phosphate buffer with electrical detection at 0.6 V (vs. Ag/AgCl) and 1218 V/cm as the applied voltage (Fig. 9.13). Gong and Hauser [33] reported cnantiomeric separation of (1S,2S)-(1)- and (1R,2R)-(2)-*trans*-cyclohexane-1,2-diamine by NCE with an effective separation length of 75 mm. The BGE used was 0.5 M acetic acid with 5 mM of DM-β-cyclodextrin and 5 mM of (+)-18-crown-6-tetracarboxilic acid (pH 2.45) with electrokinetic injection at 1 kV for 1 second.

9.4. MECHANISMS OF CHIRAL SEPARATION

The chiral recognition mechanisms in NLC and NCE devices are similar to conventional liquid chromatography and capillary electrophoresis with chiral mobile phase additives. It is important to note here that, to date, no chiral stationary phase has been developed in microfluidic devices. As discussed above polysaccharides, cyclodextrins, macrocyclic glycopeptide antibiotics, proteins, crown ethers, ligand exchangers, and Pirkle's type molecules are the most commonly used chiral selectors. These compounds have chiral cavities, grooves, baskets, and loops. On injection of racemic mixtures, the enantiomers form transient diastereoisomeric complexes with the chiral mobile phase additive. These complexes are stabilized by hydrogen bondings, van der Waals forces, π-π complexation, dipole interactions, anionic and cationic bindings, and steric forces. The distereoisomeric inclusion complexes thus formed have different physicochemical properties and, therefore, have different adsorption/partition between the mobile and stationary phases of liquid chromatography, which are the basis of their resolution. Similarly, these transient diastereoisomeric complexes have different electrophoretic properties and elute at different time intervals in NCE. For more details on chiral recognition mechanisms readers should consult our earlier book on chiral chromatography [12].

9.5. CONCLUSION

Chiral resolution is an emerging area in drug design and environmental monitoring. The analyses of drugs and xenobiotics at nano or low levels are becoming more essential and common. Consequently, NLC and NCE are of great importance in these areas. Basically, the dose of any racemic can be reduced if only one enantiomer is administered. Similarly, the exact toxicities of chiral pollutants can be ascertained by determining their enantiomeric compositions, which are of low concentrations in the environment. The demand for nano or low level detection analytical methods is therefore increasing in the present century. NLC and NCE are very important techniques in chiral drug design and environmental monitoring. However, the microfluidic separation devices are still under development and certainly will be the choice of the future in chiral analyses of biological and environmental samples in the near future.

REFERENCES

1. R.J. Haüy, *Tableaux comparatif des resultats de la crystallographie et de làanalyse chimique relativement a la classification des mineraux*, Paris (1809).
2. L. Pasteur, *C. R. Acad. Sci.*, **26**, 535 (1848).
3. G. Blaschke, H.P. Kraft, K. Fickentscher, F. Köhler, *Drug Res.*, **29**, 1640, (1979).
4. B. Knoche, G. Blaschke, *J. Chromatogr. A*, **666**, 235 (1994).
5. B. Testa, *Trends Pharmacol. Sci.*, **7**, 60 (1986).
6. R.A. Aitken, D. Parker, R.J. Taylor, J. Gopal, R.N. Kilenyi, *Asymmetric synthesis*, New York: Blackie (1992).

7. D.T. Witte, K. Ensing, J.P. Franke, R.A. de Zeeuw, *Pharm. World Sci.*, 15, 10 (1993).

8. A.G. Rauws, K. Groen, *Chirality*, **6**, 72 (1994).

9. R. Kallenborn, H. Hühnerfuss, W.A. Köning, *Angew. Chem.*, **103**, 328 (1991).

10. J. Faller, H. Hühnerfuss, W.A. Köning, R. Krebber, P. Ludwig, *Environ. Sci. Technol.*, **25**, 676 (1991).

11. I. Ali, H.Y. Aboul-Enein, *Chiral pollutants, distribution, toxicity and analysis by chromatography and capillary electrophoresis*, Chichester: Wiley (2004).

12. H.Y. Aboul-Enein, I. Ali, *Chiral separations by liquid chromatography and related technologies*, New York: Marcel Dekker (2003).

13. B. Chankvetadze, *Capillary electrophoresis in chiral analysis*, New York: Wiley (1997).

14. G. Gübitz, M.G. Schmidt (Eds.), *Chiral separations: Methods and protocol*, Totowa, NJ: Humana Press (2004).

15. R.L. Stefan, J.F. van Staden, H.Y. Aboul-Enein, *Electrochemical sensors in bioanalysis*, New York: Marcel Dekker (2001).

16. M. Pumera, *Electrophoreis*, **28**, 2113 (2007).

17. M. Pumera, I. Jelinek, J. Jindrich, O. Benada, *J. Liq. Chromatogr. Relat. Technol.*, **25**, 2473 (2002).

18. S.R. Wallenborg, I.S. Lurie, D.W. Arnold, C.G. Bailey, *Electrophoresis*, **21**, 3257 (2000).

19. I. Rodriguez, L.J. Jin, S.F.Y. Li, *Electrophoresis*, **21**, 211 (2000).

20. H.L. Zeng, H. Li, X. Wang, J.M. Lin, *J. Capil. Electrophor. Microchip Technol.*, **10**, 19 (2007).

21. T. Vickner, D. Janesek, A. Manz, *Anal. Chem.*, **76**, 3373 (2004).

22. P.S. Dittrick, K. Tachikawa, A. Manz, *Anal. Chem.*, **78**, 3887 (2006).

23. K.B. Male, J.H. Luong, *J. Chromatogr. A*, **1003**, 167 (2003).

24. M.A. Schwarz, P.C. Hauser, *J. Chromatogr. A*, **928**, 225 (2001).

25. M. Ludwig, F. Kohler, D. Belder, *Electrophoresis*, **24**, 3233 (2003)

26. K.M. Balss, W.N. Vreeland, K.W. Phinney, D. Ross, *Anal. Chem.*, **76**, 7243 (2004).

27. Y. Gao, Z. Shen, H. Wang, Z. Dai, B. Lin, *Electrophoresis*, **26**, 4774 (2005).

28. N. Piehl, M. Ludwig, D. Belder, *Electrophoresis*, **25**, 3848 (2004).

29. L.D. Hutt, D.P. Glavin, J.L. Bada, R.A. Mathies, *Anal. Chem.*, **71**, 4000 (1999).

30. H.L. Zeng, H.F. Li, X. Wang, J.M. Lin, *Talanta*, **69**, 226 (2006).

31. H. Nakajima, K. Kawata, H. Shen, T. Nakagama, K. Uchiyama, *Anal. Sci.*, **21**, 67 (2005).

32. X. Weng, H. Bi, B. Liu, J. Kong, *Electrophoresis*, **27**, 3129 (2006).

33. X.Y. Gong, P.C. Hauser, *Electrophoresis*, **27**, 4375 (2006).

CHAPTER 10

PERSPECTIVES ON NANOANALYSES

10.1. INTRODUCTION

The discovery of semiconductor integrated circuits by Bardeen, Brattain, Shockley, Kilby, and Noyce was a revolution in the micro and nano worlds. The concept of miniaturization and integration has been exploited in many areas with remarkable achievements in computers and information technology. The utility of microchips was also realized by analytical scientists and has been used in chromatography and capillary electrophoresis. In 1990, Manz et al. [1] used microfluidic devices in separation science. Later on, other scientists also worked with these units for separation and identification of various compounds. A proliferation of papers has been reported since 1990 and today a good number of publications are available in the literature on NLC and NCE. We have searched the literature through analytical and chemical abstracts, Medline, Science Finder, and peer reviewed journals and found a few thousand papers on chips but we selected only those papers related to NLC and NCE techniques. Attempts have been made to record the development of microfluidic devices in separation science. The number of papers published in the last decade (1998–2007) is shown in Fig. 10.1, which clearly indicates rapid development in microfluidic devices as analytical tools. About 30 papers were published in 1998; that number has risen to ~400 in

Nanochromatography and Nanocapillary Electrophoresis. By Ali, Aboul-Enein, and Gupta
Copyright © 2009 John Wiley & Sons, Inc.

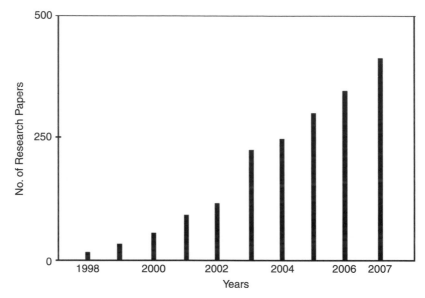

Figure 10.1 Number of research publications on NLC and NCE.

2007. This data indicates that the future of microfluidic devices in separation science and technology is quite good. The importance of nano separations has already been discussed at various places in different chapters. The chapter describes the future of nanoanalyses in separation science.

10.2. FUTURE OF MICROFLUIDIC DEVICES

Basically, microfluidic devices involve the flow of liquid in the nanoliter range, and, hence they are useful devices in separation science at nano or low level analyses of various ingredients in biological and environmental matrices. The most important applications of micro-fluidic devices include medical, chemical, and separation sciences.

The medical applications of microchips are as infusion pumps, for pain therapy, treatment of cancer and tinnitus, and in proteomics, genomics, and determining dosage of medicines. Miniaturization has enabled the development of versatile portable equipment for point-of-care monitoring and treatment of patients. In addition, microchip-based equipment has given patients choices to be more directly involved in monitoring their own health, increasing the probabilities of improving their lives and reducing health care expenses.

Drug discovery is a long, labor- and cost-intensive process and many potential drug candidates are being discovered by the use of combinatorial chemistry

[2]. Microfluidic devices are useful, fast, and inexpensive tools for studying the interactions of many drugs. Many drug discoveries and disease identification are dependent on genetic analyses, which require microfluidic devices that are capable of increasing the speed and efficiency of genetic information. The speed of these devices aids in identifying illness and quickly and efficiently treating the patient.

The chemical applications include reaction mechanisms, kinetics, and dynamics. Microchips have also gained reputation in separation science and, as discussed in earlier chapters, NLC and NCE have a wide range of applications for analyses of various molecules of a biological and environmental nature. NLC can also be integrated with other on-chip chemical and biological sensors, eliminating the need for highly specific sensing devices. Moreover, microchips have the potential of achieving high throughput separations, which can speed up research in all areas, including complex samples of biological and environmental interest. According to Robbins-Roth [3], multimillion compound libraries have been generated with synthesis at the rate of over one million new compounds per year. Therefore, conventional chromatographic and capillary electrophoretic methods cannot meet the demands and requirements of analyses of such a huge number of compounds.

Mijatovic et al. [4] reviewed the advantages and disadvantages of microfluidic devices along with their future prospects. Dittrich and Manz [5] discussed the role of microfluidics in drug discovery in terms of molecular reactions, laws of scale for surface per volume, molecular diffusion, and heat transport. Furthermore, the authors reviewed current and future applications of microfluidics and highlighted their potential in drug discovery. Whitesides [6] reviewed the origins and the future of microfluidics as potential influence in chemical synthesis, biological analysis, optics, and information technology. Chin et al. [7] reviewed past and future aspects of microfluidic devices for global health. The authors identified diseases and discussed special design criteria for these devices to be deployed in a variety of applications. The main focus was on diagnostics, and past and future aspects of these devices. Pennathur et al. [8] described the exploitation of the features of microfluidic technology and discussed their future trends. Riviere [9] imagined the role of microfluidics in veterinary therapeutics in 2030. Recently, Manz [10] wrote an editorial in *Lab-on-a-Chip Journal* describing the utility and future of microchips in various areas, from space to medical research.

10.3. FUTURE CHALLENGES

The last decade has been credited with major developments in microfluidic devices as many workers have reported on this subject. It is important to mention that microfluidic technology is not fully developed but still needs more research into nanoanalytic techniques. Therefore, some future challenges in the area are awaiting scientists to tackle them. The most critical challenges are creation of devices that are capable of performing multiple, complex fluid-handling steps in series on a fully integrated device. The sample, detection, chip interfacing, chromatographic and capillary electrophoretic separations and total integration are the issues that need to be addressed. Among these, the integration of sample preparation into microfluidic devices is one of the main hurdles to achieving true nanoanalyses. It is more challenging when analyses are to be performed on real-world samples, that is, environmental and biological matrices. Due to inbuilt characteristics, chip-based machines need nanoliter samples, which is a difficult task in conventional-scale fluidics to handle such a small amount.

The microfluidic chips are small and thin, which make most conventional coupling techniques unrealistic. In NLC, reliable coupling with high pressure ratings is highly desired, which is underdeveloped, due to the complexity of individual devices involved and the difficulty of integration for a complete on-chip system. The selectivity is usually accomplished through chromatography-based separation prior to detection and it is another problem in NLC. On the other hand, in NCE it is highly desirable and useful to know the operating conditions of the chip in real time. Besides, surface modification of NCE capillaries is the most crucial and challenging issue to achieve acceptable EOF for fast and successful separations. So far, most of the developments in nanoanalyses have been focused on discrete devices, such as microchannels, microvalves, micropumps, flow sensors, pressure sensors, and electrochemical sensors. However, total integration is an urgent and important requirement in this area. The integration of sample preparation, injection, separation channel, and detection is very much required. This is the most challenging job for scientists in the coming years. The dimensions of all the channels on microchips being used in NLC and NCE are in micro ranges, which have very good separation capabilities but we can assume the analytical potentials of chips having the dimension of channels in nano ranges. Therefore, it may be possible to fabricate nanofluidic devices for NLC and NCE. No doubt, this sort of fabrication is difficult and a challenge.

10.4. CONCLUSION

Microfluidic devices are a miracle since they have greatly reduced cost and time, making our lives easier. Even today, the chips are small and inexpensive, providing better sensitivity, selectivity, and reproducibility. However, microfluidic devices are not yet fully developed and still under development, which need a truly versatile and powerful technology. The development of nanofluidic devices is also required for more fine and low concentration analyses. In the near future, chip-based instruments will be available and useful in field tests, networked sensing, real-time environmental monitoring, and medical choices for patient treatment within minutes or seconds. Furthermore, continuous research and development of NLC and NCE instruments will solve many mysteries in genomics and proteomics. In conclusion, chip-based devices will be available, in the near future, which will have an impact on the advancement of separation science.

REFERENCES

1. A. Manz, N. Graber, H.M. Widmer, *Sensors & Actuat. B Chem.*, **1**, 244 (1990).
2. H.J. Crabtree, M. Finot, J.J. Lukomsky, V. Walker, *Lab. Chip*, **1**, 30N (2001).
3. C. Robbins-Roth, *The business of biotechnology*, Cambridge: Perseus Publishing (2000).
4. D. Mijatovic, J.C. Eijkel, A. van den Berg, *Lab. Chip*, **5**, 492 (2005).
5. P.S. Dittrich, A. Manz, *Nat. Rev. Drug. Discov.*, **5**, 210 (2006).
6. G.M. Whitesides, *Nature*, **442**, 368 (2006).
7. C.D. Chin, V. Linder, S.K. Sia, *Lab. Chip*, **7**, 41 (2007).
8. S. Pennathur, C.D. Meinhart, H.T. Soh, *Lab. Chip*, **8**, 20 (2008).
9. J.E. Riviere, *Vet. J.*, **174**, 462 (2007).
10. A. Manz, *Lab. Chip*, **8**, 13 (2008).

SUBJECT INDEX

Nanochromatography and Nanocapillary Electrophoresis. By Ali, Aboul-Enein, and Gupta
Copyright © 2009 John Wiley & Sons, Inc.